Atomically Precise
Metal Nanoclusters

Synthesis Lectures on Materials and Optics

Atomically Precise Metal Nanoclusters

Zhikun Wu and Rongchao Jin

ISBN: 978-3-031-01261-7   paperback
ISBN: 978-3-031-02789-7   ebook
ISBN: 978-3-031-00253-3   hardcover

DOI 10.1007/978-3-031-02789-7

A Publication in the Springer series
SYNTHESIS LECTURES ON MATERIALS AND OPTICS

Lecture #8
Series ISSN
Synthesis Lectures on Materials and Optics
Print 2691-1930   Electronic 2691-1949

Atomically Precise Metal Nanoclusters

Zhikun Wu
Institute of Solid State Physics, Chinese Academy of Sciences

Rongchao Jin
Carnegie Mellon University

SYNTHESIS LECTURES ON MATERIALS AND OPTICS #8

ABSTRACT

Atomically precise metal nanocluster research has emerged as a new frontier. This book serves as an introduction to metal nanoclusters protected by ligands. The authors have summarized the synthesis principles and methods, the characterization methods and new physicochemical properties, and some potential applications. By pursuing atomic precision, such nanocluster materials provide unprecedented opportunities for establishing precise relationships between the atomic-level structures and the properties. The book should be accessible to senior undergraduate and graduate students, researchers in various fields (e.g., chemistry, physics, materials, biomedicine, and engineering), R&D scientists, and science policy makers.

KEYWORDS

nanoparticles, nanoclusters, spectroscopy, microscopy, catalysis, biomedicine, energy

Contents

CHAPTER 1

Introduction

Since the beginning of the 21st century, nanoscience and nanotechnology have advanced significantly. A variety of nanostructures and devices based on such nanostructures have been achieved. Beside the basic science research, the "nano" revolution is also reflected in many industries. For example, in the computer and electronics industry, the transistor size has approached the sub-ten nanometer regime, which is a significant push to the nanotechnology development. Similarly, the chemical industry utilized a variety of nanocatalysts to speed up chemical conversions (e.g., oil reforming, methanol synthesis) and to protect the environments (e.g., three-way auto converters for eliminating emissions).

Metal nanoparticles have been extensively researched in current nanoscience and nanotechnology owing to their attractive optical, catalytic, and biological properties and a wide range of applications [1]. Among the different metal nanoparticles (e.g., noble metals and other transition metals), gold nanoparticles (NPs) are particularly appealing and there has been a long history of scientific research on such nanoparticles. Gold is the most noble metal and thus possesses a high stability, (e.g., never tarnishes in air). Gold and its nanoparticle form exhibit brilliant colors (Fig. 1.1), particularly exciting is the tunable color of gold NPs by particle size and shape (Fig. 1.1). Back in the medieval days, gold NPs were already used for the coloration of ceramics and glasses, and one such example is the famous Lycurgus Cup made by the Romans in the 4th century A.D.

1.1 FROM POLYDISPERSE NANOPARTICLES TO MONODISPERSE NANOPARTICLES

Scientific research on colloidal gold dates back to Faraday's time, who in 1857 reported a systematic study on the deep red colloidal gold (called the "potable gold") which was obtained from the reduction of aqueous $AuCl_4^-$ by phosphorus in carbon disulfide (CS_2) [2]. This seminal work initiated the research on colloids, but the origin of the wine red color of gold colloids (5–30 nm, Fig. 1.1b inset) remained mysterious, although Faraday insightfully commented that those particles of gold should be of dimensions smaller than the wavelength of visible light [2]. With the coming of 20th century, extensive research on inorganic colloids (e.g., metals such as Au, Ag, Pt, Pd, and nonmetals such as sulfur) was conducted. In particular, gold colloids attracted much attention both in experiment and theory. In 1908, Mie solved the Maxwell's equations for spherical particles and successfully modeled the extinction spectra of colloidal gold [3], but the physical mechanism of the ruby red color of gold colloids still remained unknown for many

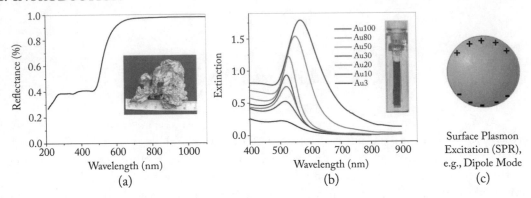

Figure 1.1: (a) Reflectance spectrum of bulk gold, (b) extinction spectra of gold colloids (3–100 nm diameter), and (c) surface plasmon excitation (e.g., dipole mode) in metal nanoparticles. Reproduced with permission from [1]. Copyright 2016 American Chemical Society.

decades more, because the electron theory of metals was still in its infancy stage and not mature yet; recall that the electron was just discovered in 1897.

During the 1900–1930s and afterward, continuous advances were achieved in the formation mechanism of inorganic colloids, their physico-chemical properties, and practical applications. It is worth noting that, during the early 20th century, some important concepts were developed, including the nucleation and growth [4], electric double layer [5], stabilization through charges [5], surface adsorption of ions [6], and the Derjaguin–Landau–Vervey–Overbeek (DLVO) theory [7–9]. The sizes of colloidal particles were primarily determined by the ultramicroscopy method (a dark-field optical microscopy technique) [10], X-ray diffraction [11], and the ultracentrifugation method [12], since the transmission electron microscopy (TEM) was not invented yet.

After the invention of TEM by Ruska and Knoll in 1932 (Nobel Prize in 1986), TEM gradually became a popular method for determing the colloid size [13–15]. The application of gold colloids in biological labeling [16] and the discovery of the dramatic enhancement of Raman scattering signals by electrochemically roughened metal surfaces [17–19] greatly stimulated the study of colloidal gold and silver NPs. During the 1950–1970s, the discovery and scientific understanding of collective electron excitation mode (as opposed to single electron excitation mode) finally explained the fundamental origin of the wine red color of gold colloids [20]; Fig. 1.1c. When light is incident on Au NPs, the surface plasmon resonance (SPR) occurs at ~520 nm wavelength for relatively small Au NPs (Fig. 1.1c), and the absorption of blue (due to interband transition) and SPR in the green (for NPs ≤ 30 nm, Fig. 1.1b) leave only the red light undisturbed, hence, a red color of the gold colloid (≤ 30 nm). For relatively small Ag NPs, the SPR at ~400 nm leads to optical absorption in the blue, leaving green and red undisturbed and mixing of these two primary colors giving rise to a yellow color of Ag colloids.

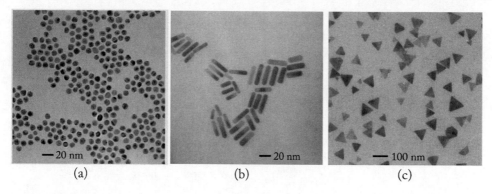

Figure 1.2: TEM images of Au and Ag NPs of different shapes. Reproduced with permission from [1]. Copyright 2016 American Chemical Society.

In the late 1990s, the era of nanotechnology started and the research on metal NPs and many other types of nanomaterials has achieved huge advances owing to significant investment in R&D of nanotechnology. Back in Faraday's time, his colloidal particles were apparently quite polydisperse. Since the beginning of the 21st century, scientists have accomplished excellent control of metal nanoparticle size and shape (Fig. 1.2) [21–24], for example, colloidal NPs with very tight size distribution (e.g., standard deviation ~5%) were already made in the past decade [1].

1.2 FROM MONODISPERSE NANOPARTICLES TO ATOMICALLY PRECISE NANOCLUSTERS

While scientists have achieved excellent control over nanoparticle size and shape, such particles are only uniform at the nanometer level (for example, 10.0 ± 1.5 nm diameter with a 15% standard deviation), but not at the atomic level. The NPs with polydispersity of 15% or less (down to 5% possibly) are typically called monodisperse samples. In most cases, such NPs are good enough and do not pose any issues. However, in some fundamental research work (e.g., surface catalysis, electron transfer, and biological interactions), researchers are still frustrated by the fact that no two NPs in a pot are the same at the atomic level, which preclude deep understanding of catalytic mechanisms and other fundamental events.

The polydispersity of NPs (though merely 5%), together with other factors (e.g., unknown surface composition and structure), poses major challenges to atomic level studies on the structure-property correlations, for instance, the catalytic mechanism for reactions on NPs, the interactions of NPs with proteins in biological systems, the interparticle electron transport mechanism, the surface trap state properties, and the surface magnetism [1]. All these call for atomically well-defined NPs and new breakthroughs are urgently needed.

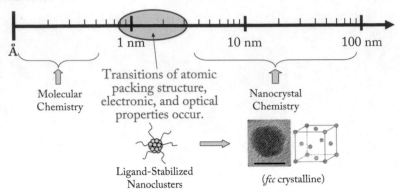

Figure 1.3: Atomically precise metal nanoclusters bridge up organometallic complexes and regular NPs (or nanocrystals). Reproduced with permission from [25]. Copyright 2010 Royal Society of Chemistry.

Recent advances have led to success in the synthesis of atomically precise NPs in the 1–3 nm size regime (often called nanoclusters, NCs); Fig. 1.3 [25]. Starting from the small sizes such as the 1 nm core $Au_{25}(SR)_{18}$ NC (where, SR refers to thiolate ligands), research has steadily progressed toward larger sizes such as the 1.7 nm core $Au_{144}(SR)_{60}$ and 2.2 nm core $Au_{246}(SR)_{80}$ [26–28]. The atomic precision and molecular purity of such unprecedented NCs have allowed researchers to successfully solve the total structures (metal core plus surface ligands) by single crystal X-ray diffraction (i.e., X-ray crystallography). Figure 1.4 shows the currently largest total structure of gold NCs [28]. It is worth noting that TEM, scanning probe microscopies (e.g., scanning tunneling microscopy) and other high-resolution techniques cannot analyze the surface/interface with atomic resolution, thus, X-ray crystallography is currently the most appealing method for characterizing the total structure of NCs.

1.3   ABOUT THIS BOOK

In this book, the synthetic methods are summarized in Chapter 2. Compared to regular NPs, the characterization of atomically precise NCs can be performed by an array of molecular characterization tools such as mass spectrometry (MS), nuclear magnetic resonance (NMR), electron paramagnetic resonance (EPR), infrared, and Raman spectroscopies. Chapter 3 illustrates the characterization of NCs by those methods, from which new properties can be learned. In contrast to the plasmonic metal NPs, the ultrasmall size of NCs leads to strong quantum size effects, which result in discrete electronic structure and molecular-like properties, such as HOMO–LUMO[1] electronic transition, enhanced photoluminescence, uique catalytic activity, redox properties, and many others (see Chapter 3). These properties are fundamentally different

[1]HOMO: highest occupied molecular orbital; LUMO: lowest unoccupied molecular orbital.

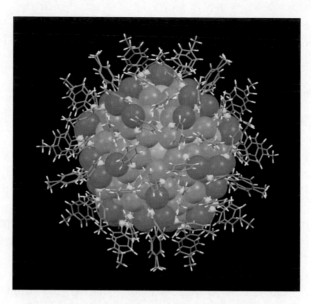

Figure 1.4: Experimental total structure of the $Au_{246}(SR)_{80}$ nanocluster (2.2 nm metal core) determined by X-ray crystallography. Reproduced with permission from [28]. Copyright 2016 AAAS.

from those of the larger counterparts—crystalline metal NPs (also called nanocrystals) in which the optical properties are dominated by plasmon excitation. The properties of NCs are very sensitive to the number of atoms in the core and the packing modes of atoms, whereas the SPR of gold NPs is much less sensitive to size, e.g., it only shifts by $\sim$50 nm when the NP size increases from $\sim$10 to $\sim$100 nm [25], which is obviously not sensitive to a single atom alteration. On the basis of the above facts, the NCs are of great significance not only for fundametal research but also for practical application [29–31]. Chapter 4 illustrates some applications of metal NCs, such as catalysis and sensing.

Overall, atomically precise NC research has emerged as an exciting frontier in recently years and is expected to flourish in the future.

1.4 REFERENCES

[1] Jin, R., Zeng, C., Zhou, M., and Chen, Y. Atomically precise colloidal metal nanoclusters and nanoparticles: Fundamentals and opportunities. *Chem. Rev.*, 116:10346–10413, 2016. DOI: 10.1021/acs.chemrev.5b00703. 1, 2, 3

[2] Faraday, M. The Bakerian lecture: Experimental relations of gold (and other metals) to light. *Philos. Transactions on R. Soc.*, 147:145–181, London 1857. DOI: 10.1098/rstl.1857.0011. 1

[3] Mie, G. Beiträge zur optik trüber medien, speziell kolloidaler metallösungen. *Ann. Phys.*, 330:377–445, 1908. DOI: 10.1002/andp.19083300302. 1

[4] Von Weimarn, P. The precipitation laws. *Chem. Rev.*, 2:217–242, 1925. DOI: 10.1021/cr60006a002. 2

[5] Wilson, J. A. Theory of colloids. *J. Am. Chem. Soc.*, 38:1982–1985, 1916. DOI: 10.1021/ja02267a009. 2

[6] Beans, H. T. and Eastlack, H. E. The electrical synthesis of colloids. *J. Am. Chem. Soc.*, 37:2667–2683, 1915. DOI: 10.1021/ja02177a010. 2

[7] Derjaguin, B. A theory of interaction of particles in presence of electric double layers and the stability of lyophobe colloids and disperse systems. *Acta Phys. Chim.*, 10:333–346, 1939. DOI: 10.1016/0079-6816(93)90010-s. 2

[8] Derjaguin, B. and Landau, L. D. Theory of the stability of strongly charged lyophobic sols and of the adhesion of strongly charged particles in solutions of electrolytes. *Acta Phys. Chim.*, 14:633–662, 1941. DOI: 10.1016/0079-6816(93)90013-l. 2

[9] Verwey, E. J. W. and Overbeek, J. T. G. *Theory of the Stability of Lyophobic Colloids*, Elsevier, Amsterdam, 1948. DOI: 10.1021/j150453a001. 2

[10] Zsigmondy, R. *Colloids and the Ultramicroscope, a Manual of Colloid Chemistry and Ultramicroscopy*, John Wiley & Sons, New York, 1909. 2

[11] Scherrer, P. Determination of the size and internal structure of colloidal particles using X-rays. *Nachr. Ges. Wiss. Goettingen. Math.-Phys. Kl.*, pages 98–100, 1918. 2

[12] Svedberg, T. and Pedersen, K. O. *The Ultracentrifuge*, Elsevier, Amsterdam, 1940. DOI: 10.1021/ac50119a001. 2

[13] Turkevich, J. and Hillier, J. Electron microscopy of colloidal systems. *Anal. Chem.*, 21:475–485, 1949. DOI: 10.1021/ac60028a009. 2

[14] Turkevich, J., Stevenson, P. C., and Hillier, J. A study of the nucleation and growth processes in the synthesis of colloidal gold. *Discuss. Faraday Soc.*, 11:55–75, 1951. DOI: 10.1039/df9511100055. 2

[15] Turkevich, J. Colloidal gold. Part I. *Gold Bull.*, 18:86–91, 1985. DOI: 10.1007/bf03214690. 2

[16] Sadler, P. *Biochemistry, Structure and Bonding*, vol. 29, Springer, Berlin, 1976. 2

[17] Fleischmann, M., Hendra, P. J., and McQuillan, A. J. Raman spectra of pyridine adsorbed at a silver electrode. *Chem. Phys. Lett.*, 26:163–166, 1974. DOI: 10.1016/0009-2614(74)85388-1. 2

[18] Albrecht, M. G. and Creighton, J. A. Anomalously intense Raman spectra of pyridine at a silver electrode. *J. Am. Chem. Soc.*, 99:5215–5217, 1977. DOI: 10.1021/ja00457a071. 2

[19] Jeanmaire, D. L. and Van Duyne, R. P. Surface Raman spectroelectrochemistry. *J. Electroanal. Chem. Interfacial Electrochem.*, 84:1–20, 1977. DOI: 10.1016/s0022-0728(77)80224-6. 2

[20] Kreibig, U. and Vollmer, M. *Optical Properties of Metal Clusters*, Springer-Verlag, New York, 1995. DOI: 10.1007/978-3-662-09109-8. 2

[21] Jin, R., Cao, Y. W., Hao, E., Metraux, G. S., Schatz, G. C., and Mirkin, C. A. Controlling anisotropic nanoparticle growth through plasmon excitation. *Nature*, 425:487–490, 2003. DOI: 10.1038/nature02020. 3

[22] Gole, A. and Murphy, C. J. Seed-mediated synthesis of gold nanorods: Role of the size and nature of the seed. *Chem. Mater.*, 16:3633–3640, 2004. DOI: 10.1021/cm0492336. 3

[23] Yang, X., Yang, M., Pang, B., Vara, M., and Xia, Y. Gold nanomaterials at work in biomedicine. *Chem. Rev.*, 115:10410–10488, 2015. DOI: 10.1021/acs.chemrev.5b00193. 3

[24] Langille, M. R., Zhang, J., Personick, M. L., Li, S., and Mirkin, C. A. Stepwise evolution of spherical seeds into 20-fold twinned icosahedra. *Science*, 337:954–957, 2012. DOI: 10.1126/science.1225653. 3

[25] Jin, R. Quantum sized, thiolate-protected gold nanoclusters. *Nanoscale*, 2:343–362, 2010. DOI: 10.1039/b9nr00160c. 4, 5

[26] Zhu, M., Aikens, C. M., Hollander, F. J., Schatz, G. C., and Jin, R. Correlating the crystal structure of a thiol-protected Au_{25} cluster and optical properties. *J. Am. Chem. Soc.*, 130:5883–5885, 2008. DOI: 10.1021/ja801173r. 4

[27] Yan, N., Xia, N., Liao, L., Zhu, M., Jin, F., Jin, R., and Wu, W. Unraveling the long-pursued Au_{144} structure by X-ray crystallography. *Science Advances*, 4:eaat7259, 2018. DOI: 10.1126/sciadv.aat7259. 4

[28] Zeng, C., Chen, Y., Kirschbaum, K., Lambright, K. J., and Jin, R. Emergence of hierarchical structural complexities in nanoparticles and their assembly. *Science*, 354:1580–1584, 2016. DOI: 10.1126/science.aak9750. 4, 5

[29] Chakraborty, I. and Pradeep, T. Atomically precise clusters of noble metals: Emerging link between atoms and nanoparticles. *Chem. Rev.*, 117:8208–8271, 2017. DOI: 10.1021/acs.chemrev.6b00769. 5

[30] Li, G. and Jin, R. Atomically precise gold nanoclusters as new model catalysts. *Acc. Chem. Res.*, 46:1749–1758, 2013. DOI: 10.1021/ar300213z. 5

[31] Du, Y., Sheng, H., Astruc, D., and Zhu, M. Atomically precise noble metal nanoclusters as efficient catalysts: A bridge between structure and properties. *Chem. Rev.*, 120:526–622, 2020. DOI: 10.1021/acs.chemrev.8b00726. 5

CHAPTER 2

Synthesis of Atomically Precise Metal Nanoclusters

2.1 INTRODUCTION

The synthesis of high-quality NPs is of paramount importance. In early research, phosphines and carbonyl were employed as the ligands to protect metal NCs and several well-defined NCs were successfully synthesized [1–5], such as Au_{11}, Au_{13}, and $Au_{13}Ag_{12}$ with crystal structures solved (see the excellent reviews by Mingos [6] and Konishi [7]). However, the phosphine-passivated Au clusters are labile and researchers have to look for alternative protecting ligands.

Inspired by the research on self-assembled monolayers of thiols on bulk gold surfaces [8], Brust et al. in 1994 employed thiols to protect gold NPs and obtained relatively stable gold NPs with 1–3 nm size due to the formation of strong Au-S bonds between thiolates and gold atoms [9]. Whetten et al. then developed a synthetic protocol for NCs and observed several stability "islands" [10, 11], although true monodispersity was not attained for those species in kDa masses.

In early work, most synthetic procedures suffered from the product dispersity and impurity [12]; therefore, the products had to be purified by employing techniques such as fractional crystallization [10], size-exclusion chromatograph (SEC) [13, 14], and gel electrophoresis [15, 16]. The difficulties in purification and low yields for the products stimulated the development of high-yield synthesis strategies.

An important advance arrived in 2008. Zhu et al. synthesized $Au_{25}(SC_2H_4Ph)_{18}$ in 40% yield (Au atom basis) [17]. The synthesis mainly consists of two steps: (i) forming the Au(I)-SR complexes by the reduction of Au(III) (e.g., $HAuCl_4$) to Au(I) with thiols and (ii) further reducing Au(I) to Au(0) by $NaBH_4$ (Fig. 2.1). It was found that the size and structure of Au-SR intermediates play a vital role in yielding $Au_{25}(SC_2H_4Ph)_{18}$ NCs based upon investigations on different reaction conditions, including the reaction stirring speed, reaction temperature, and time.

2.2 SYNTHESIS STRATEGY

Despite the notable yield improvement [17], the "one-pot one-cluster" dream was yet to be realized. By replacing the frequently used solvent toluene with THF, Wu et al. subsequently established the single phase synthesis method and, for the first time, revealed an interesting

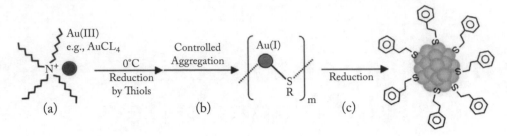

Figure 2.1: Basic steps in the synthesis of $Au_{25}(SC_2H_4Ph)_{18}$. Reproduced with permission from [17]. Copyright 2008 American Chemical Society.

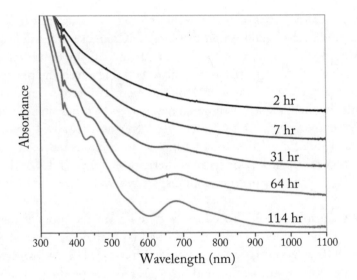

Figure 2.2: Evolution of the UV-vis spectra of the crude product with aging time. The spectra are vertically shifted for the ease of comparison. Reproduced with permission from [18]. Copyright 2009 Royal Society of Chemistry.

phenomenon—size-focusing [18], in which the polydisperse NCs are converted to monodisperse NCs of a single size. This process was demonstrated by the time dependent UV-vis spectroscopic analysis by Wu et al. [18] (Fig. 2.2) and also by matrix-assisted laser desorption ionization (MALDI) mass spectrometry analysis [19] (Fig. 2.3). By employing such a process and tuning the reaction parameters, Wu et al. obtained pure Au_{25} NCs and for the first time realized the "one-pot one cluster" dream for metal NC synthesis. The "size-focusing" process follows the basic principle of "survival of the robustest": in the harsh environment [18], those unstable NCs decompose or convert to the stable ones, and only the robustest NC can finally survive after repeated selections.

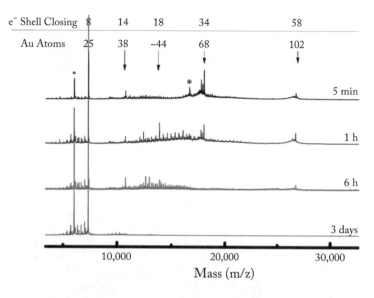

Figure 2.3: Mass spectrometric evidence of size focusing in the one-phase synthesis of $Au_{25}(SC_2H_4Ph)_{18}$ NCs. Peaks marked by asterisks are fragments. Reproduced with permission from [19]. Copyright 2009 American Chemical Society.

The size-focusing synthesis was also reported for other sizes. Qian et al. obtained $Au_{38}(SR)_{24}$ in high yield by employing the similar principle and etching a crude mixture of glutathionate-capped $Au_n(SG)_m$ NCs in a water/organic two-phase system (Fig. 2.4) [20]. The final product was dependent on the size distribution of the $Au_n(SG)_m$ precursors: the dominant precursor size range from 8–18 kDa ($\sim 38 < n < \sim 102$) results in the Au_{38} NC of high yield, while the $Au_n(SG)_m$ mixture with a dominant size range from 26–36 kDa leads to high-yielded $Au_{144}(SR)_{60}$ [22].

When the reducing reagent $NaBH_4$ was replaced by a relatively weaker reducing reagent (borane tert-butylamine complex) while keeping the other reaction conditions unchanged [18], Wu et al. synthesized a Au_{19} nanocluster instead of the Au_{25} NC. Basing on this and some other observations, Wu et al. proposed the "kinetic control and thermodynamic selection" principle for the metal NC synthesis [23].

The "kinetic control" aims to control the size distribution of the as-formed NC mixture through adjusting the reaction parameters such as stirring speed, solvent polarity, reaction temperature, reducibility of reducing agent, and so on. The "kinetic control" can influence the final product type and yield.

The "thermodynamic selection" is to create harsh environments so that the labile NCs will decompose or transform to stable ones and only the most stable NCs will finally survive, following the "survival of the robustest" rule, much like the natural law "survival of the fittest."

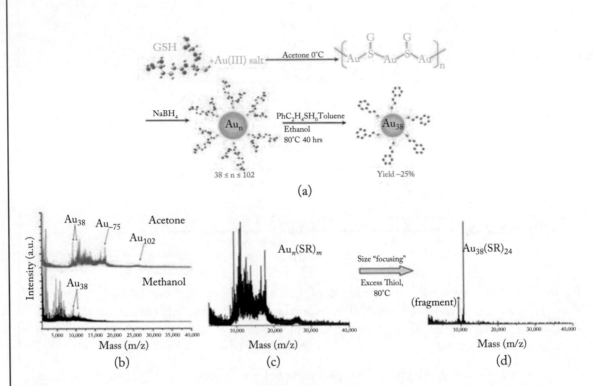

Figure 2.4: (a) Scheme of the size-focusing synthesis of $Au_{38}(SR)_{24}$ NCs; (b) the effect of solvent on the size range of the initial crude gold NCs; and (c, d) illustrate size focusing of polydisperse NCs to pure $Au_{38}(SR)_{24}$. Reproduced with permission from [20]. Copyright 2009 American Chemical Society.

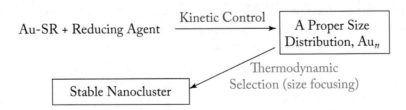

Figure 2.5: Kinetic control and thermodynamic selection in the synthesis of atomically monodisperse NCs. Reproduced with permission from [23]. Copyright 2011 American Chemical Society.

The factors that can be applied in this step include reaction temperature, time, ligand type, other assistant reagents, etc. The strategy is applicable in most of the syntheses of metal NCs. There are ample examples. Xie et al. employed carbon monoxide (CO) as a mild reducing agent and adjusted the pH value for the size-tunable syntheses of $Au_{15}(SR)_{13}$, $Au_{18}(SR)_{14}$, $Au_{25}(SR)_{18}$, and so on [24]. They also added NaOH to tune the reaction kinetics by decreasing the reducing power of $NaBH_4$ and enhancing the etching ability of thiol in the $Au_{25}(SR)_{18}$ synthesis [25]. Zeng et al. synthesized $Au_{44}(SR)_{28}$ and $Au_{52}(SR)_{32}$ under 60 and 80°C, respectively, without changing other conditions [26, 27]. Liao et al. synthesized $Au_{44}(SR)_{26}$ and $Au_{49}(SR)_{27}$, respectively, by tuning the reaction time only [28, 29]. After replacing the solvent tetrahydrofuran with dichloromethane in the synthesis of $Au_{25}(SC_2H_4Ph)_{18}$ [18], Tian et al. synthesized a low-temperature-stable $Au_{38}(SC_2H_4Ph)_{24}$ (denoted Au_{38T}) NC; this NC and the high-temperature stable $Au_{38}(SC_2H_4Ph)_{24}$ (denoted Au_{38Q}) [20] constitutes the first pair of structural isomers of metal nanoparticles [30]. Chen et al. synthesized $Au_{130}(p-MBT)_{50}$, $Au_{104}(m-MBT)_{41}$ and $Au_{40}(o-MBT)_{24}$, respectively, by simply adjusting the position of methyl group in methylbenzenethiol from *para*- to *meta*- to *ortho*-position [31]. The addition of chemical reagents is also an effective way to tune the kinetics and thermodynamics. For example, Zhuang et al. developed an acid induction method in which an acid (nitric acid, acetic acid, etc.) was added to the reaction solution after the addition of the reducing reagent $NaBH_4$. The acid can accelerate the hydrolysis of $NaBH_4$ and strengthen the reactivity of $NaBH_4$, hence accelerating the reaction kinetics [32]. However, it can weaken the interaction between Au and thiolate, reducing the reactivity of thiolate and changing the thermodynamic selection, thus influencing both the kinetics and the thermodynamics of the synthesis [32]. Several novel Au_n NCs were obtained by this method, including $Au_{52}(SC_2H_4Ph)_{32}$, $Au_{42}(SPh-t-Bu)_{26}$, $Au_{44}(SPh-t-Bu)_{26}$, and $Au_{48}(SPh-t-Bu)_{28}$ [32–34]. Further, Zhuang et al. introduced Cd^{2+} in replacement of acid and synthesized a non-fcc-structured $Au_{42}(SPh-t-Bu)_{26}$, which is the structural isomer of the existing fcc-structured $Au_{42}(SPh-t-Bu)_{26}$ (fcc: face-centered cubic) [35]. The addition of Cd(II) might influence the kinetics and thermodynamics by forming some unstable Au/Cd intermediates, tuning the $NaBH_4$ reducing ability, etc. With the establishment of the "kinetic control and thermodynamic selection" strategy, the earlier synthesis was better understood [18], and some new synthesis methods have also been developed, such as the ligand exchange—induced size/structure transformations (LEIST) [26, 27, 36] and anti-galvanic reactions [37–39]. Below we will focus on the latter two methods.

2.3   TWO MAJOR METHODS OF NANOCLUSTER CONVERSION

2.3.1   LIGAND EXCHANGE-INDUCED SIZE/STRUCTURE TRANSFORMATION (LEIST)

Early in 2007, Shichibu et al. successfully converted phosphine-protected Au_{11} to thiolated core-shell or biiosohedral Au_{25} nanoclusers by ligand-exchange [40]. While the phosphine for thiol exchange was known due to the stronger Au−SR bond than Au−PPh$_3$, the thiol for thiol exchange for new NCs indeed came as a surprise [36]. Zeng et al. reported that one thiolated gold NC can be transferred to another thiolated gold NC by ligand exchange [36]. Such a method was latter named as the ligand exchange-induced size/structure transformation (LEIST) [36]. There are two key aspects in this synthesis: the new ligand should be in great excess relative to the original ligand, and appropriate thermodynamic conditions are needed to overcome the energy barrier between the precursor NC and the product NC. This approach has an obvious advantage: the final product is typically not very complex (e.g., size mixed in a bottom-up synthesis), which favors the subsequent isolation and crystallization. Based on this size/structure conversion strategy, a series of new sizes and structures of Au NCs have been obtained, for example, $Au_{21}(SR)_{15}$ [41], $Au_{28}(SR)_{20}$ [42], $Au_{103}S_2(SR)_{41}$ [43], $Au_{133}(SR)_{52}$ [44], $Au_{279}(SR)_{84}$ [45], etc. Except for the structure transformation from non-fcc to fcc, the one-way transformation from an fcc $Au_{43}(S-c-C_6H_{11})_{25}$ to no-fcc $Au_{44}(2,4-SPhMe_2)_{26}$ ($2,4-HSPhMe_2 = 2,4-$dimethylbenzenethiol) was also reported by Dong et al. [46].

In a typical LEIST reaction, taking $Au_{38}(SC_2H_4Ph)_{24}$ as an example, the $Au_{38}(SC_2H_4Ph)_{24}$ nanocluster is first dissolved in toluene and then reacted with HSPh−t−Bu at 80°C for > 12 h. The molar ratio of incoming HSPh−t−Bu to the original protecting −SC$_2$H$_4$Ph ligand on the $Au_{38}(SC_2H_4Ph)_{24}$ surface was around 160:1. This process produced a new $Au_{36}(SPh-t-Bu)_{24}$ NC in high yield (> 90%, Au atom basis) with molecular purity [47]. The detailed transformation reaction mechanism was mapped out in this case by time-dependent mass spectrometry and UV-vis spectroscopy (Fig. 2.6). The whole process can be roughly divided into four typical stages (Fig. 2.7). The first stage involves ligand exchange without size or structure transformation, as evidenced by the fact that the optical spectra are identical to that of the starting nanoclusters. During stage II, ligand exchange continues, but the optical spectrum starts to change, indicating that the large number of bulky −SPh−t−Bu ligands on the Au_{38} core starts to induce structural distortion of the Au_{38} NC, even though the cluster formula is still $[Au_{38}(SPh-t-Bu)_n(SC_2H_4Ph)_{24-n}]$ ($n > 12$). This intermediate gives rise to a new absorption band at 550 nm in the UV-vis spectrum. The Au_{38} structural distortion preludes the subsequent size and structural transformation (stage III, 20–60 min). An interesting "disproportionation" reaction is identified, in which one Au_{38}^* cluster releases two gold atoms to form $Au_{36}(SR)_{24}$, and another $Au_{38}(SR)_{24}$ nanocluster captures the two released gold atoms and also two free −SPh−t−Bu ligands to form $Au_{40}(SR)_{26}$. This is manifested in the two new sets of mass peaks with comparable

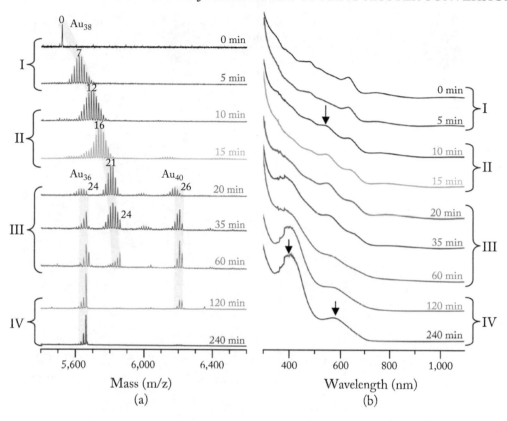

Figure 2.6: Size/structure-transformation process from $Au_{38}(SC_2H_4Ph)_{24}$ to $Au_{36}(SPh-t-Bu)_{24}$ monitored by (a) ESI-MS and (b) UV-vis. Reproduced with permission from [47]. Copyright 2013 American Chemical Society.

intensities on the lower- and higher-mass sides of the Au_{38} peak set. The lower-mass set is assigned to $Au_{36}(SPh-t-Bu)_n(SC_2H_4Ph)_{24-n}$ ($m = 19–24$), and the higher-mass set is attributed to $Au_{40}(SPh-t-Bu)_{n+2}(SC_2H_4Ph)_{24-n}$ ($n = 21–26$). As the reaction continues, Au_{38} NC is gradually converted to Au_{36} and Au_{40}. During stage IV (120–300 min), the ligand exchange comes to an end, and the $Au_{40}(SPh-t-Bu)_{n+2}(SC_2H_4Ph)_{24-n}$ species start to transform to Au_{36} under high temperature and excess thiol and eventually molecularly pure $Au_{36}(SPh-t-Bu)_{24}$ is obtained. A theoretical yield of 94% was expected on the basis of the above mechanism, and the experimental yield is 90%, which is close to the expectation, hence, supporting the mechanism [47].

It is worth noting that the ligands influence the size and structure of NCs. First, different ligands with similar hindrance might induce the formation of the same NC. For instance, the $Au_{38}(SC_2H_4Ph)_{24}$ NC can be transformed to the Au_{36} NC after etch-

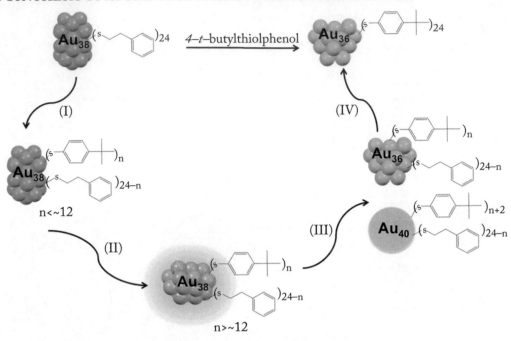

Figure 2.7: Four-stage reaction pathway for the conversion of $Au_{38}(SC_2H_4Ph)_{24}$ to $Au_{36}(SPh-t-Bu)_{24}$. Reproduced with permission from [47]. Copyright 2013 American Chemical Society.

ing with not only 4-(tert-Butyl)benzyl mercaptan but also cyclopentanethiol [48]. Gan et al. reported the synthesis of three $Au_{24}(SR)_{20}$ NCs by reacting $Au_{25}(SC_2H_4Ph)_{18}$ with different thiolate ligands ($R = C_2H_4Ph$, CH_2Ph, and $CH_2Ph-t-Bu$) [49]. Second, the same ligand might transform similar sized NCs into the identical NC; for example, both $Au_{23}(S-c-C_6H_{11})_{16}$ and $Au_{25}(SC_2H_4Ph)_{18}$ can react with $HSCH_2Ph-t-Bu$ to give rise to the same $Au_{24}(SCH_2Ph-t-Bu)_{20}$ NC [49, 50].

In addition to the ligand effect, other important influencing factors to the LEIST reaction include the incoming thiolate concentration and thermodynamic conditions. The thermodynamic conditions involve heating time, temperature, and atmosphere. As shown in the case of Au_{38} transformation to Au_{36}, an intermediate NC Au_{40} coexists at the early stage of the transformation reaction, but it finally disappears due to its less stability compared to Au_{36} under the harsh conditions [47]. By controlling the heating temperature, Zeng et al. obtained $Au_{28}(SPh-t-Bu)_{20}$ and $Au_{20}(SPh-t-Bu)_{26}$ from $Au_{25}(SC_2H_4Ph)_{18}$ under 80°C and 40°C, respectively [42, 51]. Gan et al. obtained a $Au_{60}S_6(SCH_2Ph)_{36}$ NC via a thermally induced ligand exchange reaction of molecularly pure $Au_{38}(SC_2H_4h)_{24}$ at 100°C under a nitrogen at-

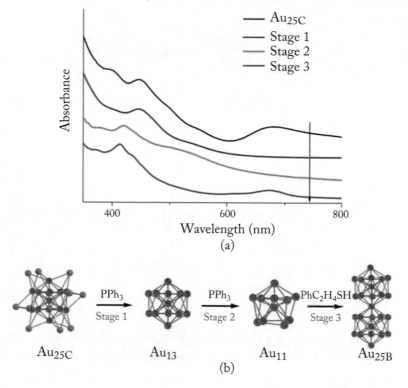

Figure 2.8: (a) UV-vis spectral evolution for the reaction between Au_{25C} and PPh_3 and (b) the tandem etching process. Reproduced with permission from [57]. Copyright 2016 American Chemical Society.

mosphere [52]. However, when the $Au_{60}S_6(SCH_2Ph)_{36}$ NC reacted with excess $HSCH_2Ph$ in air, a new NC, $Au_{60}S_7(SCH_2Ph)_{36}$, is generated [53].

Generally, strong nucleophilic ligands can be employed to exchange with the weak ones on the surface of gold core. For instance, phosphine-passivated gold NCs were transformed to thiolated gold NCs by ligand-exchange due to the fact that the Au-SR interaction is stronger than $Au-PR_3$ interaction [54–56]. The reverse process (i.e., the strong nucleophilic ligands were replaced by the weak ones) was rarely reported. Nevertheless, Li et al. reported a case of reverse ligand-exchange to tailor the composition (and structure) of Au NCs (Fig. 2.8) [57]. First, the Au_{12} exterior shell of $Au_{25}(SC_2H_4Ph)_{18}^-$ (abbreviated as Au_{25C}) was "peeled" by PPh_3; second, the remained Au_{13} core is continuously etched by PPh_3 to form $[Au_{11}(PPh_3)_8(SC_2H_4Ph)]^{2+}$ (abbreviated as Au_{11}); and finally, Au_{11} is transformed to biicosohedral $[Au_{25}(PPh_3)_{10}(SC_2H_4Ph)_5Cl_2]^{2+}$ (abbreviated as Au_{25B}). On the basis of this work, Zhu et al. investigated the reactions of PPh_3 with some thiolated gold NCs (includ-

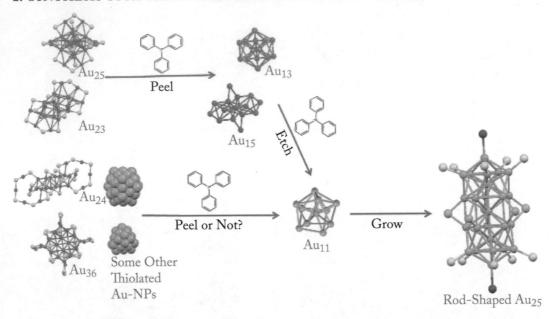

Figure 2.9: Mechanism illustration of the PPh$_3$ etching process. Reproduced with permission from [58]. Copyright 2018 Chinese Chemical Society and College of Chemistry and Molecular Engineering, Peking University.

ing Au$_{23}$(S–c–C$_6$H$_{11}$)$_{16}$, Au$_{24}$(SC$_2$H$_4$Ph)$_{20}$, Au$_{36}$(SPh–t–Bu)$_{24}$, Au$_{38}$(SC$_2$H$_4$Ph)$_{24}$), mixed NCs and ~ 3 nm Au NPs [58]. Surprisingly, the experimental results show that thiolated gold NCs (and NPs) with different compositions, structures, sizes and protecting thiolates can all be transformed to [Au$_{11}$(PPh$_3$)$_8$Cl]$^{2+}$, and finally to [Au$_{25}$(PPh$_3$)$_{10}$(SR)$_5$Cl$_2$]$^{2+}$. In other words, PPh$_3$ is a uniform "converter" for these thiolated gold NC and NPs (Fig. 2.9).

2.3.2   ANTI-GALVANIC REACTION (AGR)

Galvanic reaction, named after Italian scientist Luigi Galvani (1737–1798), involves the spontaneous reduction of a noble-metal cation by a less noble metal driven by the difference in electrochemical potentials. This reaction has been widely applied and received particular interest in nanoscience and nanotechnology [59, 60]. In 2010, Murray and co-workers first identified the species Au$_{24}$Ag(SC$_2$H$_4$Ph)$_{18}$, Au$_{23}$Ag$_2$(SC$_2$H$_4$Ph)$_{18}$, and so on, by mass spectrometry after mixing the negative Au$_{25}$(SC$_2$H$_4$Ph)$_{18}$ NC with Ag$^+$ [61]. Wu revealed that not only the negative Au$_{25}$(SC$_2$H$_4$Ph)$_{18}$ but also the neutral Au$_{25}$(SC$_2$H$_4$Ph)$_{18}$ and other ultrasmall gold and silver NPs can react with silver (or copper) ions [37, 62]. On the basis of these facts, Wu proposed the general AGR concept, that is, the reduction of metal ions by less reactive (or more noble) metals [37]. Further, Wu's group extended the AGR to Pt and Pd NPs, and demonstrated

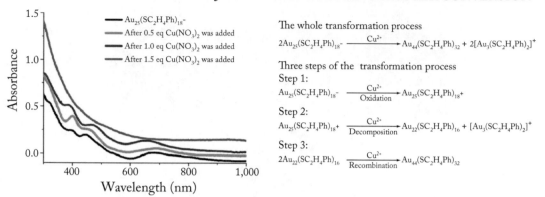

Figure 2.10: UV-Vis absorption spectral evolution of Au_{25} after the addition of Cu^{2+} and the proposed transformation process. Reproduced with permission from [62]. Copyright 2015 Royal Society of Chemistry.

that the AGR is not caused by the reducing thiolates but is intrinsically associated with the size effect of a metal, thus truly establishing the concept and started a new research topic [38, 39].

The most straightforward application of the AGR is the tuning of the composition, structure, and properties of metal NCs, and until now it is known that monometallic and alloy NCs have been synthesized by this method. The first example of the synthesis of Au NCs by way of AGR is the synthesis of $Au_{44}(SC_2H_4Ph)_{32}$ [62]. Wu's group reported that negative Au_{25} could react with Cu^{2+}, and finally produced a new product, $Au_{44}(SC_2H_4Ph)_{32}$. The whole reaction process can be divided into three stages: oxidation–reduction, decomposition, and recombination, as shown in Fig. 2.10, which was monitored by UV-Vis absorption spectra. With the addition of Cu^{2+} up to only ~ 1.0 equiv., the UV-Vis absorption spectra revealed that the anion Au_{25} was oxidized to the neutral, then to the cation as expected. With the addition of another 0.5 equiv. of Cu^{2+}, the characteristic absorption peaks of $[Au_{25}(SC_2H_4Ph)_{18}]^+$ at 400, 460, and 660 nm disappeared with the emergence of a new absorption band centered at ~ 515 nm, indicating the generation of $Au_{44}(SC_2H_4Ph)_{32}$. After this report, the same group showed that the reaction between $[Au_{23}(S-c-C_6H_{11})_{16}]^-$ and Cd^{2+} can produce $Au_{28}(S-c-C_6H_{11})_{20}$ [63].

The Wu group successfully achieved a series of doped Au NCs by use of AGR and summarized four main alloying modes of the AGR (Fig. 2.11) [63].

1. **Addition:** the foreign metal atom is introduced without altering the precursor nanocluster's composition and structure, like the doping mode which yields $Au_{25}Ag_2(SC_2H_4Ph)_{18}$ [64] or $Au_{26}Cd_5$ [65].

2. **Replacement:** the foreign metal atom substitutes the precursor metal nanocluster's metal atom with the structure framework unchanged, like the alloying mode

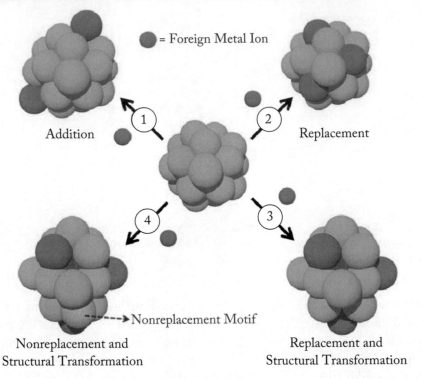

Figure 2.11: Schematic illustration of the four alloying modes by the AGR. Reproduced with permission from [63]. Copyright 2018 Wiley VCH.

which produces $Au_{24}Hg_1(SC_2H_4Ph)_{18}$ [66], $Au_{24}Cd_1(SC_2H_4Ph)_{18}$ [67], and $Au_{24-x}Ag_xHg_1(SC_2H_4Ph)_{18}$ [68].

3. **Replacement and structural transformation:** the foreign metal atom substitutes the precursor metal atom and causes the structural transformation of the precursor nanocluster, such as the alloying mode which leads to the transformation of a 23-atom alloy to 25-atom $Au_{25-x}Ag_x(SC_2H_4Ph)_{18}$ ($x \sim 19$) [69].

4. **Nonreplacement and structural transformation:** the foreign metal atom does not substitute metal atom but instead induces the structural transformation of the mother nanocluster, such as the alloying mode which leads to $Au_{20}Cd_4(SH)(S-c-C_6H_{11})_{19}$ [63].

In addition, Wu et al. found that metal nanocluster can react with the same metal complex (or salt), and they named this kind of reaction a pseudo-AGR. As an illustration, $Au_{24}(SC_2H_4Ph)_{20}$ was obtained by reacting $Au_{25}(SC_2H_4Ph)_{18}$ with the Au-PET complex (PET = $-SC_2H_4Ph$) [70]. Note that, when mixing $Au_{25}(SC_2H_4Ph)_{18}$ with $AuCl_4^-$, only large gold NPs (> 3 nm) were produced due to the stronger oxidation ability of $AuCl_4^-$ compared

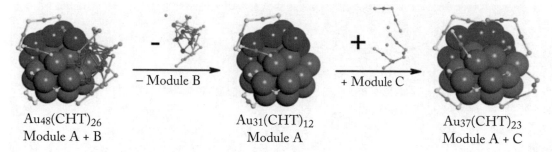

Figure 2.12: Module replacement of $Au_{48}(S-c-C_6H_{11})_{26}$ nanocluster by way of pseudo-AGR with gold cyclohexylthiolate complex. Reproduced with permission from [71]. Copyright 2020 Shanghai Institute of Organic Chemistry, Chinese Chemical Society.

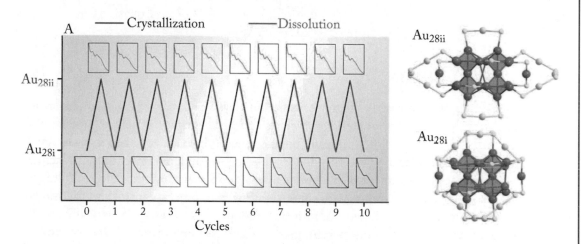

Figure 2.13: Ilustration of a pair of structural oscillators Au_{28i} and Au_{28ii} with the same composition of $Au_{28}(S-c-C_6H_{11})_{20}$. Reproduced with permission from [72]. Copyright 2020 American Chemical Society.

with that of the Au-PET complex. Employing pseudo-AGR, the same group very recently fulfilled module replacement [71] (Fig. 2.12) and also obtained a pair of structural oscillators [72] (Fig. 2.13).

Notably, Jin and co-workers reported an interesting molecular "surgery" via combination of an AGR and a pseudo-AGR (Fig. 2.14): "resection" of two surface Au atoms of $Au_{23}(SR)_{16}$ by sequentially reacting with Ag-SR and then $Au_2Cl_2(P-C-P)$ (P−C−P denotes $Ph_2PCH_2PPh_2$), leading to the formation of a new 21-gold-atom NC, $[Au_{21}(SR)_{12}(P-C-P)_2]^+$, without changing the other parts of the starting nanocluster [73].

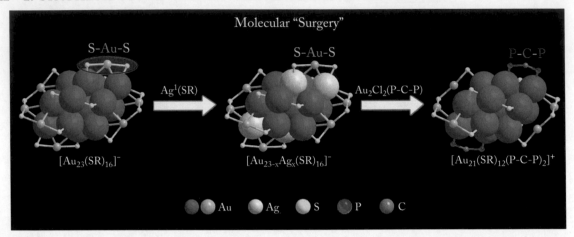

Figure 2.14: "Molecular surgery" by a combination method of AGR and quasi-AGR. Reproduced with permission from [73]. Copyright 2017 AAAS.

2.4   CONCLUSION

In materials research, the synthesis is the cornerstone and plays a crucial role in subsequent structure determination, studies of the material properties and mechanisms, as well as applications. For nanoscientists, the synthesis of atomically precise metal NPs has long been a dream goal. This goal has not been fulfilled until recently. Understanding the underlying synthesis principles is a major step for the establishment of the systematic "size-focusing" methodology. With the discovery of "size-focusing" and mechanistic understanding, the "kinetic control and thermodynamic selection" strategy was developed and implemented for the synthesis of atomically precise metal nanocluster, which offered an important guidance and also implications. Diverse synthesis methods have been deveoped until now, for example, the two major post-synthesis methods chosen for discussions in this chapter. Further understanding of the synthesis principles and the develpopment of novel synthesis methods are still needed in future work, which will certainly provide new opportunities in metal nanocluster research.

2.5   REFERENCES

[1] Schmid, G., Ed. *Clusters and Colloids*, VCH, Weinheim, Germany, 1994. DOI: 10.1002/9783527616077. 9

[2] de Silva, N. and Dahl, L. F. Synthesis and structural analysis of the first nanosized platinum—gold carbonyl/phosphine cluster, $Pt_{13}[Au_2(PPh_3)_2]_2(CO)_{10}$ $(PPh_3)_4$, containing a Pt-centered $[Ph_3PAu-AuPPh_3]$—capped icosahedral Pt_{12} cage. *Inorg. Chem.*, 44:9604–9606, 2005. DOI: 10.1021/ic050990v. 9

[3] Teo, B. K., Shi, X., and Zhang, H. Pure gold cluster of 1:9:9:1:9:9:1 layered structure: A novel 39-metal-atom cluster $[(Ph_3P)_{14}Au_{39}Cl_6]Cl_2$ with an interstitial gold atom in a hexagonal antiprismatic cage. *J. Am. Chem. Soc.*, 114:2743–2745, 1992. DOI: 10.1021/ja00033a073. 9

[4] Bertino, M. F., Sun, Z.-M., Zhang, R., and Wang, L.-S. Facile syntheses of monodisperse ultrasmall Au clusters. *J. Phys. Chem. B*, 110:21416–21418, 2006. DOI: 10.1021/jp065227g. 9

[5] Woehrle, G. H., Warner, M. G., and Hutchison, J. E. J. Ligand exchange reactions yield subnanometer, thiol-stabilized gold particles with defined optical transitions. *Phys. Chem. B*, 106:9979–9981, 2002. DOI: 10.1021/jp025943s. 9

[6] Mingos, D. M. P. Structural and bonding patterns in gold clusters. *Dalton Transactions on*, 44:6680–6695, 2015. DOI: 10.1039/c5dt00253b. 9

[7] Konishi, K. *Phosphine-coordinated pure-gold clusters: Diverse geometrical structures and unique optical properties/responses.* In *Gold Clusters, Colloids and Nanoparticles I*, Mingos, D. M. P., Ed., *Structure and Bonding*, 161:49–86, Springer International Publishing, Berlin, 2014. 9

[8] Bain, C. D., Troughton, E. B., Tao, Y. T., Evall, J., Whitesides, G. M., Nuzzo, R. G. Formation of monolayer films by the spontaneous assembly of organic thiols from solution onto gold. *J. Am. Chem. Soc.*, 111:321–335, 1989. DOI: 10.1021/ja00183a049. 9

[9] Brust, M., Walker, M., Bethell, D., Schiffrin, D. J., and Whyman, R. Synthesis of thiol-derivatized gold nanoparticles in a two-phase liquid-liquid system. *J. Chem. Soc. Chem. Commun.*, pages 801–802, 1994. DOI: 10.1039/c39940000801. 9

[10] Alvarez, M. M., Khoury, J. T., Schaaff, T. G., Shafigullin, M., Vezmar, I., and Whetten, R. L. Critical sizes in the growth of Au clusters. *Chem. Phys. Lett.*, 266:91–98, 1997. DOI: 10.1016/s0009-2614(96)01535-7. 9

[11] Wyrwas, R. B., Alvarez, M. M., Khoury, J. T., Price, R. C., Schaaff, T. G., and Whetten, R. L. The colours of nanometric gold. *Eur. Phys. J. D*, 43:91–95, 2007. DOI: 10.1140/epjd/e2007-00117-6. 9

[12] Jin, R. Quantum sized, thiolate-protected gold nanoclusters. *Nanoscale*, 2:343–362, 2010. DOI: 10.1039/b9nr00160c. 9

[13] Wilcoxon, J. P. and Provencio, P. Etching and aging effects in nanosize Au clusters investigated using high-resolution size-exclusion chromatography. *J. Phys. Chem. B*, 107:12949–12957, 2003. DOI: 10.1021/jp027575y. 9

[14] Donkers, R. L., Lee, D., and Murray, R. W. Synthesis and isolation of the molecule-like cluster $Au_{38}(PhCH_2CH_2S)_{24}$. *Langmuir*, 20:1945–1952, 2004. DOI: 10.1021/la035706w. 9

[15] Negishi, Y., Nobusada, K., and Tsukuda, T. Glutathione-protected gold clusters revisited: Bridging the gap between gold(I)—thiolate complexes and thiolate-protected gold nanocrystals. *J. Am. Chem. Soc.*, 127:5261–5270, 2005. DOI: 10.1021/ja042218h. 9

[16] Shichibu, Y., Negishi, Y., Tsunoyama, H., Kanehara, M., Teranishi, T., and Tsukuda, T. Extremely high stability of glutathionate-protected Au_{25} clusters against core etching. *Small*, 3:835–839, 2007. DOI: 10.1002/smll.200600611. 9

[17] Zhu, M., Lanni, E., Garg, N., Bier, M. E., and Jin, R. Kinetically controlled, high-yield synthesis of Au_{25} clusters. *J. Am. Chem. Soc.*, 130:1138–1139, 2008. DOI: 10.1021/ja0782448. 9, 10

[18] Wu, Z., Suhan, J., and Jin, R. One-pot synthesis of atomically monodisperse, thiol-functionalized Au_{25} nanoclusters. *J. Mater. Chem.*, 19:622–626, 2009. DOI: 10.1039/b815983a. 10, 11, 13

[19] Dharmaratne, A. C., Krick, T., and Dass, A. Nanocluster size evolution studied by mass spectrometry in room temperature $Au_{25}(SR)_{18}$ synthesis. *J. Am. Chem. Soc.*, 131:13604–13605, 2009. DOI: 10.1021/ja906087a. 10, 11

[20] Qian, H., Zhu, Y., and Jin, R. Size-focusing synthesis, optical and electrochemical properties of monodisperse $Au_{38}(SC_2H_4Ph)_{24}$ nanoclusters. *ACS Nano*, 3:3795–3803, 2009. DOI: 10.1021/nn901137h. 11, 12, 13

[21] Qian, H., Eckenhoff, W. T., Zhu, Y., Pintauer, T., and Jin, R. Total structure determination of thiolate-protected Au_{38} nanoparticles. *J. Am. Chem. Soc.*, 132:8280–8281, 2010. DOI: 10.1021/ja103592z.

[22] Qian, H. and Jin, R. Controlling nanoparticles with atomic precision: The case of $Au_{144}(SCH_2CH_2Ph)_{60}$. *Nano Lett.*, 9:4083–4087, 2009. DOI: 10.1021/nl902300y. 11

[23] Wu, Z., MacDonald, M. A., Chen, J., Zhang, P., and Jin, R. Kinetic control and thermodynamic selection in the synthesis of atomically precise gold nanoclusters. *J. Am. Chem. Soc.*, 133:9670–9673, 2011. DOI: 10.1021/ja2028102. 11, 12

[24] Yu, Y., Chen, X., Yao, Q., Yu, Y., Yan, N., and Xie, J. Scalable and precise synthesis of thiolated Au_{10-12}, Au_{15}, Au_{18}, and Au_{25} nanoclusters via pH controlled CO reduction. *Chem. Mater.*, 25:946–952, 2013. DOI: 10.1021/cm304098x. 13

[25] Yuan, X., Zhang, B., Luo, Z., Yao, Q., Leong, D. T., Yan, N., and Xie, J. Balancing the rate of cluster growth and etching for gram-scale synthesis of thiolate-protected Au_{25} nanoclusters with atomic precision. *Angew. Chem. Int. Ed.*, 53:4623–4627, 2014. DOI: 10.1002/anie.201311177. 13

[26] Zeng, C., Chen, Y., Iida, K., Nobusada, K., Kirschbaum, K., Lambright, K. J., and Jin, R. Gold quantum boxes: On the periodicities and the quantum confinement in the Au_{28}, Au_{36}, Au_{44}, and Au_{52} magic series. *J. Am. Chem. Soc.*, 138:3950–3953, 2016. DOI: 10.1021/jacs.5b12747.s001. 13

[27] Zeng, C., Chen, Y., Liu, C., Nobusada, K., Rosi, N. L., and Jin, R. Gold tetrahedra coil up: Kekulé-like and double helical superstructures. *Sci. Adv.*, 1:e1500425, 2015. DOI: 10.1126/sciadv.1500425. 13

[28] Liao, L., Zhuang, S., Wang, P., Xu, Y., Yan, N., Dong, H., Wang, C., Zhao, Y., Xia, N., Li, J., Deng, H., Pei, Y., Tian, S.-K., and Wu, Z. Quasi-dual-packed-Kerneled Au_{49}(2,4-DMBT)$_{27}$ nanoclusters and the influence of Kernel packing on the electrochemical gap. *Angew. Chem. Int. Ed.*, 56:12644–12648, 2017. DOI: 10.1002/anie.201707582. 13

[29] Liao, L., Zhuang, S., Yao, C., Yan, N., Chen, J., Wang, C., Xia, N., Liu, X., Li, M.-B., Lo, L., Bao, X., and Wu, Z. Structure of chiral Au_{44}(2,4-DMBT)$_{26}$ nanocluster with an 18-electron shell closure. *J. Am. Chem. Soc.*, 138:10425–10428, 2006. DOI: 10.1021/jacs.6b07178. 13

[30] Tian, S., Li, Y.-Z., Li, M.-B., Yuan, J., Yang, J., Wu, Z., and Jin, R. Structural isomerism in gold nanoparticles revealed by X-ray crystallography. *Nat. Commun.*, 6:8667, 2015. DOI: 10.1038/ncomms9667. 13

[31] Chen, Y., Zeng, C., Kauffman, D. R., and Jin, R. Tuning the magic size of atomically precise gold nanoclusters via isomeric methylbenzenethiols. *Nano Lett.*, 15:3603–3609, 2015. DOI: 10.1021/acs.nanolett.5b01122. 13

[32] Zhuang, S., Liao, L., Li, M.-B., Yao, C., Zhao, Y., Dong, H., Li, J., Deng, H., Li, L., and Wu, Z. The fcc structure isomerization in gold nanoclusters. *Nanoscale*, 9:14809–14813, 2017. DOI: 10.1039/c7nr05239a. 13

[33] Zhuang, S., Liao, L., Yuan, J., Wang, C., Zhao, Y., Xia, N., Gan, Z., Gu, W., Li, J., Deng, H., Yang, J., and Wu, Z. Kernel homology in gold nanoclusters. *Angew. Chem. Int. Ed.*, 57:15450–15454, 2018. DOI: 10.1002/anie.201808997. 13

[34] Zhuang, S., Liao, L., Zhao, Y., Yuan, J., Yao, C., Liu, X., Li, J., Deng, H., Yang, J., and Wu, Z. Is the kernel-staples match a key-lock match?. *Chem. Sci.*, 9:2437–2442, 2018. DOI: 10.1039/c7sc05019d. 13

[35] Zhuang, S., Liao, L., Yuan, J., Xia, N., Zhao, Y., Wang, C., Gan, Z., Yan, N., He, L., Li, J., Deng, H., Guan, Z., Yang, J., and Wu, Z. Fcc versus non-fcc structural isomerism of gold nanoparticles with kernel atom packing dependent photoluminescence. *Angew. Chem. Int. Ed.*, 58:4510–4514, 2019. DOI: 10.1002/anie.201813426. 13

[36] Zeng, C., Chen, Y., Das, A., and Jin, R. Transformation chemistry of gold nanoclusters: From one stable size to another. *J. Phys. Chem. Lett.*, 6:2976–2986, 2015. DOI: 10.1021/acs.jpclett.5b01150. 13, 14

[37] Wu, Z. Anti-galvanic reduction of thiolate-protected gold and silver nanoparticles. *Angew. Chem. Int. Ed.*, 51:2934–2938, 2012. DOI: 10.1002/anie.201107822. 13, 18

[38] Wang, M., Wu, Z., Chu, Z., Yang, J., and Yao, C. Chemico-physical synthesis of surfactant-and ligand-free gold nanoparticles and their anti-galvanic reduction property. *Chem. Asian J.*, 9:1006–1010, 2014. DOI: 10.1002/asia.201301562. 13, 19

[39] Gan, Z., Xia, N., and Wu, Z. Discovery, mechanism, and application of antigalvanic reaction. *Acc. Chem. Res.*, 51:2774–2783, 2018. DOI: 10.1021/acs.accounts.8b00374. 13, 19

[40] Shichibu, Y., Negishi, Y., Tsukuda, T., and Teranishi, T. Large-scale synthesis of thiolated Au_{25} clusters via ligand exchange reactions of phosphine-stabilized Au_{11} clusters. *J. Am. Chem. Soc.*, 127:13464–13465, 2005. DOI: 10.1021/ja053915s. 14

[41] Chen, S., Xiong, L., Wang, S., Ma, Z., Jin, S., Sheng, H., Pei, Y., and Zhu, M. Total structure determination of $Au_{21}(S\text{-}Adm)_{15}$ and geometrical/electronic structure evolution of thiolated gold nanoclusters. *J. Am. Chem. Soc.*, 138:10754–10757, 2016. DOI: 10.1021/jacs.6b06004. 14

[42] Zeng, C., Li, T., Das, A., Rosi, N. L., and Jin, R. Chiral structure of thiolate-protected 28-gold-atom nanocluster determined by X-ray crystallography. *J. Am. Chem. Soc.*, 135:10011–10013, 2013. DOI: 10.1021/ja404058q. 14, 16

[43] Higaki, T., Liu, C., Zhou, M., Luo, T.-Y., Rosi, N. L., and Jin, R. Tailoring the structure of 58-electron gold nanoclusters: $Au_{103}S_2(S\text{-}Nap)_{41}$ and its implications. *J. Am. Chem. Soc.*, 139:9994–10001, 2017. DOI: 10.1021/jacs.7b04678. 14

[44] Zeng, C., Chen, Y., Kirschbaum, K., Appavoo, K., Sfeir, M. Y., and Jin, R. Structural patterns at all scales in a nonmetallic chiral $Au_{133}(SR)_{52}$ nanoparticle. *Sci. Adv.*, 1:e1500045, 2015. DOI: 10.1126/sciadv.1500045. 14

[45] Higaki, T., Zhou, M., Lambright, K. J., Kirschbaum, K., Sfeir, M. Y., and Jin, R. Sharp transition from nonmetallic Au_{246} to metallic Au_{279} with nascent surface plasmon resonance. *J. Am. Chem. Soc.*, 140:5691–5695, 2018. DOI: 10.1021/jacs.8b02487. 14

[46] Dong, H., Liao, L., and Wu, Z. Two-way transformation between fcc- and nonfcc-structured gold nanoclusters. *J. Phys. Chem. Lett.*, 8:5338–5343, 2017. DOI: 10.1021/acs.jpclett.7b02459. 14

[47] Zeng, C., Liu, C., Pei, Y., and Jin, R. Thiol ligand-induced transformation of $Au_{38}(SC_2H_4Ph)_{24}$ to $Au_{36}(SPh-t-Bu)_{24}$. *ACS Nano*, 7:6138–6145, 2013. DOI: 10.1021/nn401971g. 14, 15, 16

[48] Das, A., Liu, C., Zeng, C., Li, G., Li, T., Rosi, N. L., and Jin, R. Cyclopentanethiolato-protected $Au_{36}(SC_5H_9)_{24}$ nanocluster: Crystal structure and implications for the steric and electronic effects of ligand. *J. Phys. Chem. A*, 118:8264–8269, 2014. DOI: 10.1021/jp501073a. 16

[49] Gan, Z., Lin, Y., Luo, L., Han, G., Liu, W., Liu, Z., Yao, C., Weng, L., Liao, L., Chen, J., Liu, X., Luo, Y., Wang, C., Wei, S., and Wu, Z. Fluorescent gold nanoclusters with interlocked staples and a fully thiolate-bound kernel. *Angew. Chem. Int. Ed.*, 55:11567–11571, 2016. DOI: 10.1002/anie.201606661. 16

[50] Das, A., Li, T., Li, G., Nobusada, K., Zeng, C., Rosi, N. L., and Jin, R. Crystal structure and electronic properties of a thiolate-protected Au_{24} nanocluster. *Nanoscale*, 6:6458–6462, 2014. DOI: 10.1039/c4nr01350f. 16

[51] Zeng, C., Liu, C., Chen, Y., Rosi, N. L., and Jin, R. Gold-thiolate ring as a protecting motif in the $Au_{20}(SR)_{16}$ nanocluster and implications. *J. Am. Chem. Soc.*, 136:11922–11925, 2014. DOI: 10.1021/ja506802n. 16

[52] Gan, Z., Chen, J., Wang, J., Wang, C., Li, M.-B., Yao, C., Zhuang, S., Xu, A., Li, L., and Wu, Z. The fourth crystallographic closest packing unveiled in the gold nanocluster crystal. *Nat. Commun.*, 8:14739, 2017. DOI: 10.1038/ncomms14739. 17

[53] Gan, Z., Chen, J., Liao, L., Zhang, H., and Wu, Z. Surface single-atom tailoring of a gold nanoparticle. *J. Phys. Chem. Lett.*, 9:204–208, 2018. DOI: 10.1021/acs.jpclett.7b02982. 17

[54] Brown, L. O. and Hutchison, J. E. Convenient preparation of stable, narrow-dispersity, gold nanocrystals by ligand exchange reactions. *J. Am. Chem. Soc.*, 119:12384–12385, 1997. DOI: 10.1021/ja972900u. 17

[55] Shichibu, Y., Negishi, Y., Tsukuda, T., and Teranishi, T. Large-scale synthesis of thiolated Au_{25} clusters via ligand exchange reactions of phosphine-stabilized Au_{11} clusters. *J. Am. Chem. Soc.*, 127:13464–13465, 2005. DOI: 10.1021/ja053915s. 17

[56] Woehrle, G. H., Brown, L. O., and Hutchison, J. E. Thiol-functionalized, 1.5-nm gold nanoparticles through ligand exchange reactions: Scope and mechanism of ligand exchange. *J. Am. Chem. Soc.*, 127:2172–2183, 2005. DOI: 10.1021/ja0457718. 17

[57] Li, M.-B., Tian, S.-K., Wu, Z., and Jin, R. Peeling the core-shell Au_{25} nanocluster by reverse ligand-exchange. *Chem. Mater.*, 28:1022–1025, 2016. DOI: 10.1021/acs.chemmater.5b04907. 17

[58] Zhu, M., Li, M., Yao, C., Xia, P., Zhao, Y., Yan, N., Liao, L., and Wu, Z. PPh_3: Converts thiolated gold nanoparticles to $Au_{25}(PPh_3)_{10}$ $(SR)_5Cl_2^{2+}$. *Acta Phys. Chim. Sin.*, 34:792–798, 2018. 18

[59] Liu, X. and Astruc, D. From galvanic to anti-galvanic synthesis of bimetallic nanoparticles and applications in catalysis, sensing, and materials science. *Adv. Mater.*, 29:1605305, 2017. DOI: 10.1002/adma.201605305. 18

[60] da Silva, A. G. M., Rodrigues, T. S., Haigh, S. J., and Camargo, P. H. C. Galvanic replacement reaction: Recent developments for engineering metal nanostructures towards catalytic applications. *Chem. Commun.*, 53:7135–7148, 2017. DOI: 10.1039/c7cc02352a. 18

[61] Choi, J.-P., Fields-Zinna, C. A., Stiles, R. L., Balasubramanian, R., Douglas, A. D., Crowe, M. C., and Murray, R. W. Reactivity of $[Au_{25}(SCH_2CH_2Ph)_{18}]^-$ nanoparticles with metal ions. *J. Phys. Chem. C*, 114:15890–15896, 2010. DOI: 10.1021/jp9101114. 18

[62] Li, M.-B., Tian, S.-K., Wu, Z., and Jin, R. Cu^{2+} induced formation of $Au_{44}(SC_2H_4Ph)_{32}$ and its high catalytic activity for the reduction of 4-nitrophenol at low temperature. *Chem. Commun.*, 51:4433–4436, 2015. DOI: 10.1039/c4cc08830a. 18, 19

[63] Zhu, M., Wang, P., Yan, N., Chai, X., He, L., Zhao, Y., Xia, N., Yao, C., Li, J., Deng, H., Zhu, Y., Pei, Y., and Wu, Z. The fourth alloying mode by way of anti-galvanic reaction. *Angew. Chem. Int. Ed.*, 57:4500–4504, 2018. DOI: 10.1002/anie.201800877. 19, 20

[64] Yao, C., Chen, J., Li, M.-B., Liu, L., Yang, J., and Wu, Z. Adding two active silver atoms on Au_{25} nanoparticle. *Nano Lett.*, 15:1281–1287, 2015. DOI: 10.1021/nl504477t. 19

[65] Li, M.-b., Tian, S.-k., and Wu, Z. Improving the catalytic activity of Au_{25} nanocluster by peeling and doping. *Chin. J. Chem.*, 35:567–571, 2017. DOI: 10.1002/cjoc.201600526. 19

[66] Liao, L., Zhou, S., Dai, Y., Liu, L., Yao, C., Fu, C., Yang, J., and Wu, Z. Mono-mercury doping of Au_{25} and the HOMO/LUMO energies evaluation employing differential pulse voltammetry. *J. Am. Chem. Soc.*, 137:9511–9514, 2015. DOI: 10.1021/jacs.5b03483. 20

[67] Yao, C., Lin, Y.-j., Yuan, J., Liao, L., Zhu, M., Weng, L.-h., Yang, J., and Wu, Z. Mono-cadmium vs. mono-mercury doping of Au_{25} nanoclusters. *J. Am. Chem. Soc.*, 137:15350–15353, 2015. DOI: 10.1021/jacs.5b09627. 20

[68] Yan, N., Liao, L. W., Yuan, J. Y., Lin, Y. J., Weng, L. H., Yang, J. L., and Wu, Z. K. Bimetal doping in nanoclusters: Synergistic or counteractive?. *Chem. Mater.*, 28:8240–8247, 2016. DOI: 10.1021/acs.chemmater.6b03132. 20

[69] Li, Q., Wang, S. X., Kirschbaum, K., Lambright, K. J., Das, A., and Jin, R. C. Heavily doped $Au_{25-x}Ag_x(SC_6H_{11})_{18}^-$ nanoclusters: Silver goes from the core to the surface. *Chem. Commun.*, 52:5194–5197, 2016. DOI: 10.1039/C6CC01243D. 20

[70] Yao, C., Tian, S., Liao, L., Liu, X., Xia, N., Yan, N., Gan, Z., and Wu, Z. Synthesis of fluorescent phenylethanethiolated gold nanoclusters via pseudo-AGR method. *Nanoscale*, 7:16200–16203, 2015. DOI: 10.1039/c5nr04760a. 20

[71] Jin, F., Dong, H., Zhao, Y., Zhuang, S., Liao, L., Yan, N., Gu, W., Zha, J., Yuan, J., Li, J., Deng, H., Gan, Z., Yang, J., and Wu, Z. Module replacement of gold nanoparticles by a pseudo-AGR process. *Acta Chim. Sinica (Huaxue Xuebao)*, 78:407–411, 2020. DOI: 10.6023/a20040134. 21

[72] Xia, N., Yuan, J., Liao, L., Zhang, W., Li, J., Deng, H., Yang, J., and Wu Z. Structural oscillation revealed in gold nanoparticles. *J. Am. Chem. Soc.*, 142:12140–12145, 2020. DOI: 10.1021/jacs.0c02117. 21

[73] Li, Q., Luo, T.-Y., Taylor, M. G., Wang, S., Zhu, X., Song, Y., Mpourmpakis, G., Rosi, N. L., and Jin, R. Molecular "surgery" on a 23-gold-atom nanoparticle. *Sci. Adv.*, 3:e1603193, 2017. DOI: 10.1126/sciadv.1603193. 21, 22

CHAPTER 3

Characterization of Atomically Precise Metal Nanoclusters

3.1 INTRODUCTION

With the development of wet-chemistry synthetic methods (e.g., direct reduction, ligand-exchange induced transformation, anti-galvanic reaction and solid state route), various atomically precise mono- and multi-metallic NCs (e.g., Au, Ag, and Cu) have been successfully synthesized [1–10]. The most striking feature of atomically precise metal NCs is that they have a distinct number of metal atoms (several to hundreds) and a specific number of protecting ligands (stabilizing the NC), with definite crystal structures. The formula of a NC can be described as M_nL_m, where M and L refer to the metal and protecting ligand, and n and m are the number of metal atoms and protecting ligands, respectively. Such a well-defined composition can be determined by mass spectrometer [11, 12]. In addition, single-crystal X-ray crystallography can be used to unravel the unique total structure of atomically precise metal NCs [1–10, 13].

The optical, electronic, and catalytic properties of metal NCs are strongly dependent on their structure, composition, and surface configuration, thus, various experimental techniques have been applied to the characterizations of metal NCs [14–16]. What's more, the detailed structural characterizations allow researchers to deeply understand the structure—property correlation, which provides the fundamental principle for NCs' structural optimization for specific applications [17]. For example, one should understand the structural features of fluorescent metal NCs by combining single-crystal X-ray diffraction (SCXD), X-ray absorption spectroscopy (XAS), and photoluminescence spectroscopy, and then provide some guidance for the synthesis of metal NCs with stronger fluorescence [18]; one can also determine the structure, composition, and redox properties by SCXD, mass spectrometry, and cyclic voltammetry, respectively, and then optimize the catalyst design in cases the electron-binding ability of NCs affects their catalytic performances [19].

Recently, a number of techniques have been applied to probe the structure, composition, and other aspects of atomically precise metal NCs. In this chapter, we summarize the common characterization techniques, including structure analysis (including SCXRD, microcrystal electron diffraction, XAS, nuclear magnetic resonance, etc.), size and composition analysis (such as mass spectrometry (MS), transmission electron microscopy (TEM), X-ray photoelectron spectroscopy (XPS), thermogravimetric analysis (TGA), etc.), studies on the optical properties (e.g., UV-vis absorption spectroscopy, photoluminescence spectroscopy, ultrafast electron dynamics, circular dichroism (CD) spectroscopy, etc.), electrochemical properties (e.g., cyclic voltammetry, differential pulse voltammetry, etc.), and magnetic properties (e.g., electron paramagnetic resonance). It should be noted that due to the tiny size of metal NCs, some methods for metal NPs (nanocrystals) characterization may not be suitable for NCs. For instance, TEM is a powerful tool to determine the size and shape of regular nanocrystals [20], but it cannot directly image the interface between the metal core and organic ligands, nor identify the precise core size (i.e., the number of atoms in the metal core) of NCs. The surface ligands of metal NCs include thiolate, selenolate, tellurolate, phosphine, alkyne, mixed ligands, etc., we mainly focus on the gold NCs protected by thiolate, which is the most widely studied.

3.2   STRUCTURE CHARACTERIZATIONS

3.2.1   SINGLE-CRYSTAL X-RAY DIFFRACTION (SCXRD)

The structure of metal NCs plays a decisive role in determining their unique properties. Therefore, precise knowledge of the NC atomic-level structure is of paramount importance to understand the atom packing modes within the NC and metal-ligand interactions, as well as the NC packing in macroscopic single crystals. Based upon the definitive crystal structure, the relationship between the structure and the electronic, optical, and catalytic properties as well as the size-dependent evolution can be ultimately understood.

SCXRD is the most reliable approach for the determination of the total structure of metal NCs [1]. In order to carry out this characterization, high-purity metal NCs must be synthesized and high-quality single crystals be grown. However, obtaining single crystal structures of metal NCs is challenging primarily because of the insufficient crystal quality due to the much larger sizes of NCs than small molecules. In addition to the NC purity, the crystallization methodology and solvent/antisolvent pair also play important roles in forming diffractable crystals. In most cases, the diffusion and evaporation methods work well, but a concentrated NC solution is required for these cases and the clusters must be highly pure. It is worth noting that Maran et al. reported an electrocrystallization strategy to prepare the crystals of NCs in large quantities and high quality [21].

A number of NC structures have been solved by SCXRD, and many fundamental issues can now be understood. In the following we highlight some representative structures. The first crystal structure of thiolated Au NCs was $Au_{102}(SR)_{44}$ (SR = thiolate ligand), reported by Kornberg and co-workers in 2007 (Fig. 3.1a) [13]. The cluster is composed of a Marks decahedral core of 49 gold atoms, two 20-atom caps with 5-fold rotational symmetry on two opposite poles, and a 13-atom equatorial band. Each sulfur atom of the ligands (p-mercaptobenzoic acid, p-MBA for short) binds to two gold atoms in a bridge mode and thus forms a rigid surface layer on the metal core (often called the "staple" motifs). The chiral nature of the cluster was also revealed due to the specific geometry of the atoms in the equatorial band.

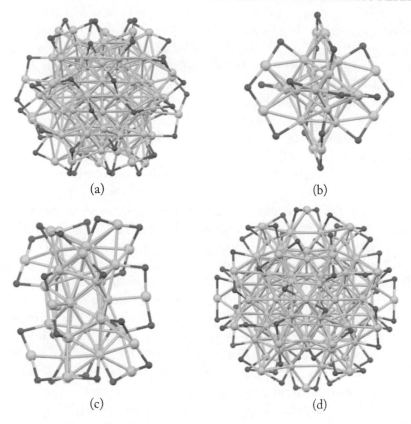

Figure 3.1: X-ray structures of (a) $Au_{102}(SR)_{44}$ [13], (b) $Au_{25}(SR)_{18}$ [22, 23], (c) $Au_{38}(SR)_{24}$ [24], and (d) $Au_{144}(SR)_{60}$ [31]. Yellow, Au; green, S. The carbon tails are omitted for clarity. Redrawn from cifs.

In 2008, the $Au_{25}(SR)_{18}$ structure was solved by the Jin and Murray groups, which possesses an icosahedral Au_{13} core and the exterior 12 Au atoms in the form of six [–S–Au–S–Au–S–] staples (Fig. 3.1b) [22, 23].

In 2010, the total structure of $Au_{38}(SR)_{24}$ was unraveled by Jin and co-workers, which showed a face-fused bi-icosahedral Au_{23} core protected by three monomeric $Au(SR)_2$ staples at the waist of the Au_{23} rod and six dimeric $Au_2(SR)_3$ staples capped at the two ends of the Au_{23} rod (three for each end) (Fig. 3.1c) [24].

Thus, the structures of several stable thiolated gold NCs were solved in the early stage (2007–2010), except for $Au_{144}(SR)_{60}$ [25]. The $Au_{144}(SR)_{60}$ NC was first reported as a 28–30 kDa species by Whetten's group in 1997, and then as $Au_{144}(SR)_{59}$ by Tsukuda's group in 2008, and finally Jin's group coined the $Au_{144}(SR)_{60}$ formula in 2009. However, for some reason, numerous attempts on crystallization of $Au_{144}(SR)_{60}$ were not successful, thus its structure had long remained mysterious [25]. Several groups claimed that they acquired single crystals of the Au_{144} NC with various ligands, but they failed in the SCXRD due to the orientation disorders of NCs in single crystals [28]. In 2016, Zeng et al. succeeded in crystallization of a giant $Au_{246}(SR)_{80}$ NC (see Chapter 1, Fig. 1.4) and the ligand interaction modes were reported to be critically important for crystallization of large-sized NCs [30]. Inspired by the success of $Au_{246}(SR)_{80}$, the long-pursued structure of $Au_{144}(SR)_{60}$ was finally solved by Yan et al. in 2018 [31]. They obtained high-quality single crystals of $Au_{144}(SR)_{60}$ (R = CH_2Ph) and successfully solved its structure by SCXRD on the basis of the rationale that the intra- and interparticle weak interactions play critical roles in growing high-quality single crystals of metal NCs (Fig. 3.1d) [31].

The crystal structure of $Au_{144}(SR)_{60}$ is made up of a Au_{114} core protected by 30 monomeric staples. The core of Au_{114} is made up of a triple shell-by-shell structure, starting from a hollow inner shell of icosahedral Au_{12}, then a middle shell of icosahedral Au_{42}, finally an outer shell of rhombicosidodecahedral Au_{60}. The surface protecting monomeric staples are on Au_4 squares of the exposed 60 Au atoms. The arrangement of staple motifs on the Au_{114} core's surface induces profound chirality to the Au_{144} NC and this NC structure exhibits a chiral I-symmetry.

It is worth noting that the gold cores in the above NCs adopt either decahedron (D_h) or icosahedron (I_h) structures, rather than the well-known face-centered cubic (fcc) one in the bulk gold and regular Au NPs. Until now, various types of gold NCs with non-fcc core structures have been reported, such as $Au_{18}(SR)_{14}$ and $Au_{30}(SR)_{18}$ (hexagonal close packed, hcp) [32–34], $Au_{20}(SR)_{16}$ and $Au_{24}(SR)_{20}$ (bitetrahedral) [35–37], $Au_{38}S_2(SR)_{20}$ (body-centered cubic, bcc) [38], $Au_{103}S_2(SR)_{41}$ (D_h) [39], $Au_{133}(SR)_{52}$ (I_h) [40], and $Au_{246}(SR)_{80}$ (D_h) [41].

The fcc $Au_n(SR)_m$ structure—which was once thought unstable in ultrasmall gold NCs—was first discovered by Zeng et al. in 2012 [42]. Subsequent work found that the fcc-structured gold NCs indeed have a large family. The $Au_{36}(SR)_{24}$ structure consists of a 28-atom fcc core with 4 interpenetrating cuboctahedrons and 4 $Au_2(SR)_3$ dimeric staples (one each (111) facet) and 12 simple –S(R)-bridges on six (100) facets [42]. Later, other fcc structures include $Au_{28}(SR)_{20}$ [43], $Au_{44}(SR)_{28}$ [44], $Au_{52}(SR)_{32}$ [45], $Au_{92}(SR)_{44}$ [46], and $Au_{279}(SR)_{84}$ [47] were reported. Interestingly, the $Au_{28}(SR)_{20}$, $Au_{36}(SR)_{24}$, $Au_{44}(SR)_{28}$, and $Au_{52}(SR)_{32}$ form a magic-size family and the kernel atoms of this magic series grows through sequential interpenetration of two additional cuboctahedral units at the bottom of the smaller kernel (Fig. 3.2) [44]. In an alternative view, each (001) layer in the kernel of the 4 NCs is a full layer with 8 gold atoms (Au_8 unit) and the molecular formula of these NCs can be summed up as a unified molecular formula $Au_{8n+4}(SR)_{4n+8}$ (n is the number of Au_8 units). Very recently, Wu's group has obtained a novel $Au_{56}(SR)_{34}$ NC with a defective kernel layer composed of 4 gold atoms (rather than 8 atoms), revealing a layer stacking mode in which a half monolayer composed of 4 atoms (Au_4 unit) is stacked on the full monolayer along the (001) direction [49].

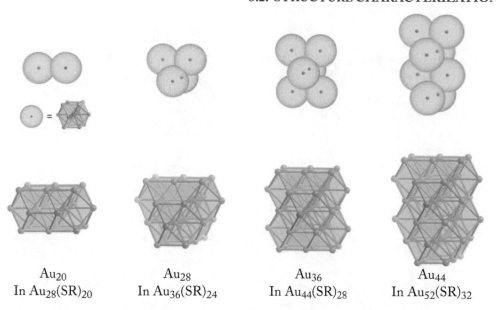

| Au$_{20}$ | Au$_{28}$ | Au$_{36}$ | Au$_{44}$ |
| In Au$_{28}$(SR)$_{20}$ | In Au$_{36}$(SR)$_{24}$ | In Au$_{44}$(SR)$_{28}$ | In Au$_{52}$(SR)$_{32}$ |

Figure 3.2: The cuboctahedron interpenetration mode in the series of NCs including Au$_{28}$(SR)$_{20}$, Au$_{36}$(SR)$_{24}$, Au$_{44}$(SR)$_{28}$, and Au$_{52}$(SR)$_{32}$. Reproduced with permission from [44]. Copyright 2016 American Chemical Society.

Except the thiols as ligands, gold NCs protected by other ligands have also been reported, with the structures unraveled, including phosphine, seleno, alkynyl, halogen, as well as mix ligands. For example, the Au$_{11}$ structure [50] led to a prediction of the Au$_{13}$ NC to possess a centered icosahedral geometry [51], which was then confirmed by experimental SCXRD analysis by Mingos et al [51]. Teo et al. reported the [Au$_{39}$(PPh$_3$)$_{14}$Cl$_6$]Cl$_2$ structure, which exhibited an hcp structure [53]. Chen et al. reported a Au$_{22}$ NC protected by six bidentate diphosphine ligands [1,8-bis(diphenylphosphino) octane (dppo)] [54]. Tsukuda's group obtained [Au$_{25}$(PPh$_3$)$_{10}$(SC$_2$H$_5$)$_5$Cl$_2$]$^{2+}$ through thiol etching of phosphine-protected Au$_{11}$ and revealed the structure to be a biicosahedral rod, with the five thiolate ligands joining the two icosahedra [55]. The Zhu group reported the structure of a Au$_{24}$(SePh)$_{20}$ NC, consisting of a prolate Au$_8$ kernel which was different from the Au$_{24}$(SR)$_{20}$ NC [56]. The Wang group crystallized several alkynyl-propected NCs such as [Au$_{19}$L$_9$(PPR)$_3$]$^{2+}$, [Au$_{23}$L$_9$(PR)$_6$]$^{2+}$, and [Au$_{24}$L$_{14}$(PR)$_4$]$^{2+}$ (L = PhC≡C-) [57].

The formation of diffractable crystals and determination of the structures are difficult for silver NCs due to their less stability. Nevertheless, there have been notable advances in recent years and some structures have been reported by, for instance, the Bigioni, Zheng, Pradeep, Bakr, Zhu, and other groups [58–62]. Yang et al. reported the crystal structure of a mixed ligand protected Ag$_{14}$(SC$_6$H$_3$F$_2$)$_{12}$(PPh$_3$)$_8$ NC in 2012 [63]. The Ag$_{14}$ cluster contains an oc-

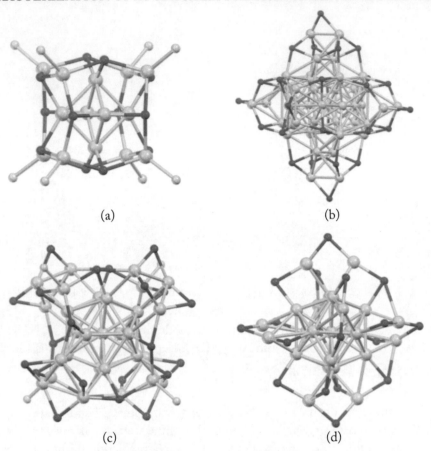

(a)   (b)

(c)   (d)

Figure 3.3: X-ray structures of (a) Ag_{14} [63], (b) Ag_{44} [64], (c) Ag_{29} [66], and (d) Ag_{25} [67]. Blue, Ag; green, S; pink, P. The carbon tails are omitted for clarity. Redrawn from the cifs.

tahedral Ag_6^{4+} core which is encapsulated by eight cubically arranged $[Ag^+(SC_6H_3F_2^-)_2PPh_3]$ units (Fig. 3.3a). Crystal structures of the Ag_{44} NC protected with p-mercaptobenzoic acid and p-fluorothiophenol, respectively, have also been solved [64, 65]. The hollow cage containing 12 silver atoms arranged in an icosahedron is surrounded by 20 silver atoms. This $Ag_{12}@Ag_{20}$ core is further encapsulated with six $Ag_2(SR)_5$ units octahedrally (Fig. 3.3b), which is different from the staples in thiolated Au NCs. The crystal structure of a dithiol-protected silver NC, $Ag_{29}(BDT)_{12}(TPP)_4$ (where, BDT = benzenedithiol; TPP = triphenylphosphine) has been solved by Bakr's group, which possesses an Ag_{13} icosahedral core protected with a shell of $Ag_{16}S_{24}P_4$ (Fig. 3.3c) [66]. The "golden silver" $Ag_{25}(SR)_{18}$ structure was also reported by the same group, which has a structure similar to that of $Au_{25}(SR)_{18}$, that is, an Ag_{13} icosahedral core surrounded by six $Ag_2(SR)_3$ staple motifs (Fig. 3.3d) [67]. Structural studies of Ag NCs have been summarized in some reviews [1, 2, 8, 58, 68].

There are also structural identifications of other metals such as copper and palladium. In 2015, Hayton's group first reported the synthesis of $[Cu_{25}H_{22}(PPh_3)_{12}]Cl$ with a Cu_{13} centered-icosahedron capped by four triangular $[Cu(PPh_3)]_3$ units, which are arranged at the corners of a tetrahedron [69]. It is noteworthy that this is the first copper NC with partial Cu (0) character. Dahl et al. reported the structures of I_h $Pd_{145}(CO)_x(PEt_3)_{30}$ [70]. Recently, they also characterized the structure of a Mackay icosahedral $Pd_{55}(PR_3)_{12}(CO)_{20}$ NC [71].

The crystal structures discussed above are all about mono-metallic NCs. With regards to multi-metallic NCs, they can be classified into two categories. One category is that the atom(s) of an existing NC are replaced by atom(s) of another metal, which can be called "doped." For example, $Au_{24}Hg(SR)_{18}$, $Au_{24}Cd(SR)_{18}$, $Au_{25-x}Ag_x(SR)_{18}$ and $Au_{24-x}HgAg_x(SR)_{18}$ have the same structures as that of $Au_{25}(SR)_{18}$ [72–75], but the doped atoms replace gold atoms in different positions. The other category is that the parent cluster is not existent, which can be called "alloyed." For instance, $[Ag_{28}Cu_{12}(SR)_{24}]^{4-}$ is an alloy cluster, with the parent $[Ag_{40}(SR)_{24}]^{4-}$ cluster unknown, at least by now [76]. These two types of bi- and multi-metallic NCs have been widely reported with definite structures, which greatly expand the knowledge of the metal atom packing form and property evolution.

3.2.2   MICROCRYSTAL ELECTRON DIFFRACTION (MicroED)

An important prerequisite of SCXRD is that high-quality single crystals of metal NCs are required. However, metal NCs protected by aqueous ligands are usually difficult to grow into large enough single crystals due to the long-chain or bulky ligand, solubility, and other complications. To overcome this barrier, electron crystallography [77], which has been used to identify the structure of small organic molecules and proteins, has been introduced for the structure characterization of aqueous metal NCs.

Electron crystallography is similar to X-ray crystallography in that a crystal scatters a beam to produce a diffraction pattern. However, the interactions between the electron beam and the crystal are much stronger than those between the X-ray photons and the crystal. This implies that a larger amount of data can be collected from much smaller crystals. Vergara et al. solved the structure of $Au_{146}(p-MBA)_{57}$ NC by MicroED at subatomic resolution (0.85 Å) and by X-ray diffraction at atomic resolution (1.3 Å) [78]. Note that the subatomic resolution was only achieved with MicroED where crystals were analyzed in a frozen hydrated state by electron diffraction (Fig. 3.4a). A total of 146 gold atoms and 57 p-MBA ligands (Fig. 3.4b) were identified in the electron density maps together with mass spectrometry. The 146 gold atoms may be decomposed into 2 sets: 119 core atoms with a twinned fcc structure and 27 peripheral atoms which follow a C_2 rotational symmetry bisected the twinning plane. The protecting layer of 57 p-MBA ligands fully encloses the core and comprises different motifs, including bridging, monomeric, and dimeric staples.

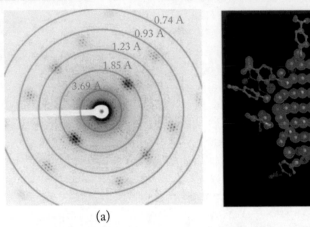

(a)   (b)

Figure 3.4: (a) Typical MicroED data of $Au_{146}(p-MBA)_{57}$ extending beyond 1 Å; and (b) MicroED density map, shown as red mesh, identifies atomic positions of Au (white spheres) and S (yellow spheres) atoms. Ligands (p-MBA) are shown as blue frameworks. Reproduced with permission from [78]. Copyright 2017 American Chemical Society.

3.2.3   X-RAY ABSORPTION SPECTROSCOPY (XAS)

When the ligand-protected metal NCs are non-crystalline, or stabilized on supported materials such as carbon and polymer, the SCXRD method cannot be used to identify their structures, even though their sizes and compositions are precisely defined. In this regard, XAS is a versatile and powerful method for determining the local structure and electronic state of a specific element within the NC regardless of whether the NCs are in dispersion or an amorphous solid.

A typical XAS spectrum is obtained by measuring absorbance as a function of X-ray energy and consists of two spectral regions, X-ray absorption near edge structure (XANES) and extended X-ray absorption fine structure (EXAFS) [79]. In regard to the application in metal NC structure characterization, for instance, holes in the d state can be determined by monitoring the intensity of the first spectral feature in XANES spectra, and specific information on the local structure of the X-ray-absorbing atom (e.g., bond length, coordination number) can be obtained from EXAFS signals. A series of works have been carried out using the XAS techniques by the Zhang, Tsukuda, Wei and some other groups [79–81]. Herein we highlight some important results.

In terms of structural identification, the first XAS study of atomically precise Au NCs was conducted on Au_{144} by MacDonald et al. [82]. The structural information from EXAFS fitting analysis was found to be consistent with the DFT model by Lopez–Acevedo and co-workers [83]. Omoda et al. investigated the structure of $[Au_{25}(MPG)_{18}]^-$ (MPG = N-(2-mercaptopropionyl)glycine) by EXAFS analysis and density functional theory (DFT) calculations [84]. EXAFS oscillation of $[Au_{25}(MPG)_{18}]^-$ differs significantly from that of

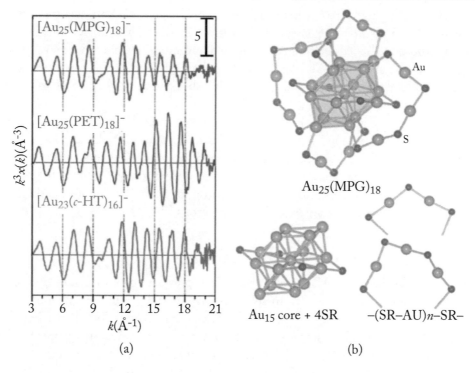

Figure 3.5: (a) Au L_3-edge EXAFS oscillations at 10 K of different Au NCs and (b) structure model of $Au_{25}(MPG)_{18}$. Reproduced with permission from [84]. Copyright 2018 American Chemical Society.

$[Au_{25}(PET)_{18}]^-$ (PET = phenylethanethiol), but is similar to that of $[Au_{23}(c-HT)_{16}]^-$ (c-HT = cyclohexanethiol) having a fcc structured core (Fig. 3.5a). They proposed a model structure, that is, an Au_{15} core with an fcc structure which is protected by four bridging thiolates and two types of staples with different lengths, $Au_2(MPG)_3$ and $Au_3(MPG)_4$ (Fig. 3.5b). The Wei group reported an icosahedron-to-cuboctahedron structural transformation of Au NCs driven by a changing chemical environment, in which the icosahedral Au_{13} clusters protected by binary ligands (dodecanethiolate and triphenylphosphine) can be converted to a cuboctahedral structure when the solvent was changed from ethanol to hexane, in which the introduced solvent led to the rapid selective desorption of the thiolates [81].

In variable temperature testing, MacDonald et al. reported that a decrease of bond length is observed within the Au_{13} icosahedral kernel upon temperature decrease, but the Au (kernel)-Au (staple) bond length expands, implying metal-like behavior of the kernel and molecule-like behavior of the staple shell [85]. The Tsukuda group investigated bond stiffness in $Au_{25}(SR)_{18}$, $Au_{38}(SR)_{24}$, and $Au_{144}(SR)_{60}$ by temperature-dependent EXAFS spectroscopy [86]. They found that the Au-S bonds are much stiffer than the Au-Au bonds and there are two types of Au-Au bonds; one is stiffer and the other is softer than those in the bulk Au. Specifically, the long Au-Au bonds (those at the icosahedral core's surface) are more flexible than those in the bulk metal. In contrast, the short Au-Au bonds (in the radial direction of the core) are stiffer than those in the bulk metal and form a cyclic structural backbone with rigid Au-S oligomeric staple motifs [86].

In terms of doping atom locations, the Zhang group used Pt and Au L_3-edge EXAFS to investigate the mono-Pt doped $Au_{24}Pt$ NC and the location of Pt dopant was determined to be at the NC center [87]. Tsukuda's group examined the occupying site of Pd dopant in the $Au_{24}Pd$ NC and indicated that the single Pd atom is also preferentially located in the center [88]. Subsequent works by Wu and Ackerson groups confirm the center-doping of Pt and Pd by using SCXRD and nuclear magnetic resonance characterizations [89, 90].

3.2.4   NUCLEAR MAGNETIC RESONANCE (NMR)

Nuclear magnetic resonance (NMR) spectroscopy is a powerful tool in chemical sciences for the identification of organic molecules in solution. Based on the different chemical shifts of surface ligands (e.g., those on staples) and metal core-bound ligands, NMR has been used to reveal the surface structure (symmetry) of metal NCs in solution and investigate the dynamic nature of surface ligands, which are otherwise difficulty to obtain by characterizing the solid crystal [91, 92]. For example, Wu et al. studied the effect of different types of thiolate ligands on the Au_{25} NCs by using 1D and 2D NMR. They found that the $Au_{25}(SR)_{18}$ NC (SR = glutathionate, hexylthiolate, and dodecanethiolate) adopted the same core-shell structure of the crystallographically characterized $Au_{25}(SC_2H_4Ph)_{18}$ NC, indicating the high structural stability of $Au_{25}(SR)_{18}$ NCs [93]. In order to probe the packing structure of the carbon groups of the $Au_{130}(p-MBT)_{50}$ NC (p-MBT = 4-methylbenzenethiolate) which were not fully resolved in the SCXRD analysis, Chen et al. employed ^{1}H-NMR and found that the 50 p-MBT ligands are split into 5 groups, with each group containing 10 identical ligands (Fig. 3.6) [94]. Such a 5-fold splitting pattern of carbon groups indeed correlates with the D_5 symmetry of the surface structure in the $Au_{130}S_{50}$ framework [94]. Xia et al. employed ^{1}H-NMR to probe the nature of the staple motif on the $Ag_{30}(Capt)_{18}$ NC surface (Capt = captopril), and three types of chemically distinct thiolate–silver binding modes were found [95]. A combination of different staple types was proposed, such as a combination of $-[RS_a-Ag-S_aR]-$ and $-[RS_a-Ag-S_bR-Ag-S_aR]-$ staples [95].

Furthermore, NMR spectroscopy can also provide important information on the structural stability, charge state, magnetism, and chirality of metal NCs. The Jin group monitored the changes of NMR spectra of $Au_{25}(SG)_{18}$ (SG = glutathione) over the course of oxidation and heating [96]. Both experiments explicitly demonstrated the distinct stability differences between the two binding modes of surface ligands. The binding mode in which the ligands link to the Au_{13} core showed more stability than the binding mode in which the ligands locate in

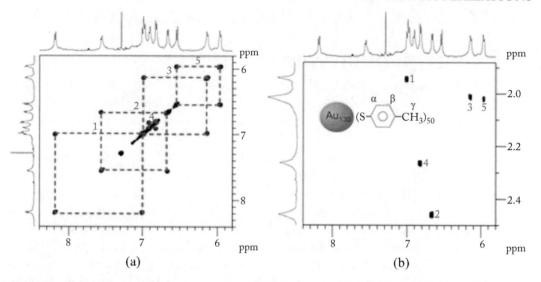

Figure 3.6: Two-dimensional NMR of $Au_{130}(p-MBT)_{50}$: (a) $^1H-^1H$ COSY of α-H and β-H and (b) $^1H-^1H$ COSY of β-H and γ-H. Reproduced with permission from [94]. Copyright 2015 American Chemical Society.

the middle of –S–Au–S–Au–S– staples [96]. Murray and co-workers described how the ^1H-NMR changes upon oxidation of $Au_{25}(SC_2H_4Ph)_{18}$ [97]. It was demonstrated that this NC can change its oxidation state from −1 to 0 and even +1, much like a "simple" molecular redox species [97]. Maran and co-workers reported a study of combined DFT and NMR, in which they explained how the different oxidation states of Au_{25} affect the ^1H- and ^{13}C-NMR signals [98]. Zeng et al. compared the NMR spectra of $[Au_{133}(TBBT)_{52}]^0$ and $[Au_{133}(TBBT)_{52}]^+$ (TBBT = 4-*tert*-butylbenzenethiolate). When paramagnetic $[Au_{133}(TBBT)_{52}]^0$ is oxidized to diamagnetic $[Au_{133}(TBBT)_{52}]^+$, the peaks of NMR spectrum shift to both lower and higher fields, which is resulted from the change of magnetic state and charge state of the gold core [99]. Jin and co-workers reported an NMR study of the intrinsically chiral $Au_{38}(SC_2H_4Ph)_{24}$ NC [100]. The chirality of $Au_{38}(SC_2H_4Ph)_{24}$ originates from the dual-propeller-like distribution of the six $Au_2(SR)_3$ dimeric staples on the Au_{23} rod-shaped core (Fig. 3.1c, Section 2.1). The $Au_{38}(SC_2H_4Ph)_{24}$ NC is found to exhibit a large chemical shift difference in the α-CH_2 protons of different ligands caused by the chiral distribution of the staples on the cluster surface [100].

3.3   SIZE AND COMPOSITION CHARACTERIZATIONS

3.3.1   MASS SPECTROMETRY (MS)

As one of the most important traditional analytical techniques, MS has been widely used to provide information on the molecular weight of a sample by the mass/charge ratio and the charge number. Moreover, MS can analyze the chemical structures of organic compounds and small biological molecules [11, 12].

Since the 1990s, MS has been successfully applied to the determination of chemical compositions of gold NCs by Whetten and others (Fig. 3.7) [25, 101]. MS is considered as a powerful tool for the size (metal atom number) and composition characterizations of various types of metal NCs. Careful studies on high-resolution MS from Murray's and Tsukuda's groups, among others, have unraveled the molecular formulas of metal NCs [102, 103]. Besides, MS has also been used to probe the charge states and study the growth mechanism of metal NCs. There are two commonly used MS methods in NC studies: matrix-assisted laser desorption ionization MS (MALDI-MS) and electrospray ionization MS (ESI-MS). Each of these two methods has its own advantages, which will be detailed in the following discussions.

In early work on NCs, the laser desorption ionization MS (LDI-MS) without matrix assistance was usually employed but could not determine the exact composition of the NCs [101]. The Whetten group first found the application of LDI-MS in gold NCs. In their work, the crude product with mixed sizes was first separated through cycles of fractional crystallization from solution and then monitored at each stage by LDI-TOF-MS analysis with negative ion mode. The mass gradually changed with the separated fractions varied, indicating the effective size determination of metal NCs from LDI-MS measurements [25, 101]. In another study, Tsukuda and co-workers prepared thiolate-protected gold NCs by thiol exchange of poly(N-vinyl-2-pyrrodidone)-stabilized Au NCs. After separating with gel permeation chromatography (GPC), the fraction containing only 11 kDa species was assigned as $Au_{55}(SC_{18})_{32}$ NC (~ 1.3 nm) through the LDI-MS characterization [103]. These results indicate that LDI-MS can be used as an efficient and direct method to identify the size (number of metal atoms and the surface ligands) and purity of metal NCs.

However, the laser irradiation in LDI-MS can cleave the S–C and Au–S bonds or even cause the loss of gold atoms. To remedy this problem, Dass et al. used trans-2-[3-(4-*tert*-butylphenyl)-2-methyl-2-propenyldidene] malononitrile (DCTB) as matrix, which ionizes NCs by electron transfer rather than the proton transfer of weak organic acid matrices, and obtained the dominant peak around 7391 Da that corresponds to the molecular ion of the $Au_{25}(SC_2H_4Ph)_{18}$ NC, indicating that the ligands remain intact on the gold NC core (Fig. 3.8) [104]. Following this report, several groups have shown the great advantage of MALDI-MS using DCTB as the matrix [105–107]. Nowadays, MALDI-MS has been successfully applied to the analysis of large sized NCs (up to approximately m/z of 400 kDa, ~ 2000 Au atoms) and the purity identification [108].

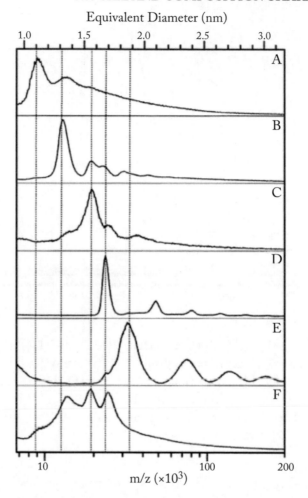

Figure 3.7: LDI-mass spectra of various magic-sized NPs. Spectra (a)–(e) each corresponds to a particular magic size, having been fractionated from a mixture of synthesized NPs. Spectrum (f) is a representative unfractionated mixture of NPs with dodecanethiolate ligands. Reproduced with permission from [101]. Copyright 1997 American Chemical Society.

Compared to MALDI-MS, ESI-MS is a much softer ionization technique that tends to produce more abundant molecular ions and little or no fragmentation. Thus, the application of ESI-MS in metal NC analysis allows one to determine the exact mass/charge ratio and composition of metal NCs. The Whetten group and subsequently Tsukuda group successfully applied ESI-MS in the studies of small glutathionate-protected gold NCs [109–112]. For example, Tsukuda and co-workers studied the $Au_n(SG)_m$ system carefully and analyzed all the separated

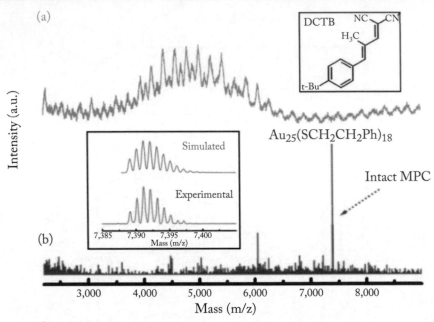

Figure 3.8: (a) Positive MALDI-TOF-MS spectrum of $Au_{25}(SC_2H_4Ph)_{18}$ with HABA (4'-hydroxyazobenzene-2-carboxylic acid) matrix (gray). (b) Positive MALDI-TOF-MS spectrum of $Au_{25}(SC_2H_4Ph)_{18}$ with DCTB matrix (red, see inset). Inset shows the expansion of isotopic resolution of the molecular ion peak. Blue spectrum is simulated, and red spectrum is the experimental data, shifted by two mass units due to experimental error. Reproduced with permission from [104]. Copyright 2008 American Chemical Society.

cluster species with ESI-MS. First, $Au_n(SG)_m$ NCs were freshly synthesized and then fractionated into nine NC components by high resolution polyacrylamide gel electrophoresis (PAGE). Then, the isolated NC species were analyzed by ESI-MS and were assigned as $Au_{10}(SG)_{10}$, $Au_{15}(SG)_{13}$, $Au_{18}(SG)_{14}$, $Au_{22}(SG)_{16}$, $Au_{22}(SG)_{17}$, $Au_{25}(SG)_{18}$, $Au_{29}(SG)_{20}$, $Au_{33}(SG)_{22}$, and $Au_{39}(SG)_{24}$. In this study, the $Au_{25}(SG)_{18}$ nanocluser was for the first time correctly identified [112].

Although ESI-MS works well with aqueous soluble gold NCs as they are easily ionizable, for organic soluble NCs, this technique is not well suited. To overcome these problems, researchers have shown that ionization efficiency can be improved by adding external ionizing agents such as CsOAc. So far, the largest atomically precise Au NC identified by ESI-MS is $Au_{333}(SR)_{79}$ [113].

Except for gold, significant efforts of MS have also been devoted to the characterization of Ag, Cu, Pt, and Pd NCs or complexes. Jin's group succeeded in ESI-MS analysis of Ag_7 NC for the first time [114]. On the basis of the dominant peak at m/z 1520.4 in the mass spectrum, the formula of silver NCs was determined to be $Ag_7(DMSA)_4$, where DMSA is 2,3-dimercaptosuccinic acid (Fig. 3.9) [114]. The $Ag_{44}(SR)_{30}$ NC was identified by ESI-MS [64]. The Chen group synthesized copper NCs protected by 2-mercapto-5-n-propylpyrimidine and analyzed the composition by positive-ion ESI-MS [115]. The result showed that Cu_8 clusters are the dominant component in the colloid solution [115]. Inouye and co-workers synthesized Pt_5 NCs according to the ESI-MS analysis [116]. Chen et al. identified a family of $[Pd(SC_2H_4Ph)_2]_n$ NCs (17 sizes total, $n = 4$–20) using MALDI-TOF-MS [117]. Moreover, the composition characterization of multimetallic NCs is also carried out by employing MS. Negishi et al. successfully synthesized AuAg bimetallic NCs and all the peaks in the MALDI-MS spectra can be assigned to a series of $Au_{25-n}Ag_n(SC_{12}H_{25})_{18}$ ($n = 0$–11) bimetallic NCs [118]. Qian et al. reported the characterization of $Pt_1Au_{24}(SR)_{18}$ NCs [119]. The Wu group successfully synthesized and analyzed the formula of $Au_{25}Ag_2(SC_2H_4Ph)_{18}$ NC [120].

MS can also be used to identify the charge states and study the growth mechanism of metal NCs. Tsukuda and co-workers [121] discussed the charge states of $[Au_{25}(SC_6H_{13})_{18}]^q$ ($q = -1, 0, +1$) on the basis of detailed ESI-MS analysis. A single peak assigned to $[Au_{25}(SC_6H_{13})_{18}]^+$ was observed in the positive-ion ESI mass spectra. In contrast, only a single $[Au_{25}(SC_6H_{13})_{18}]^-$ species was detected in the negative-ion ESI mass spectrum [121]. Murray's group reported the use of alkali metal salt for adduct formation to determine the charge states of $Au_{25}(SR)_{18}$ NCs ($-1, 0, +1, +2$) [122]. Dass and co-workers monitored the formation progress of Au_{25} NC by carrying out a MS investigation and confirmed the "size-focusing" process proposed by Wu et al. [123, 124]. Jin's group studied the size/structure transformation from $Au_{38}(PET)_{24}$ to $Au_{36}(TBBT)_{24}$ monitored by ESI-MS and optical measurements [125]. Recently, the Xie group developed time-course ESI-MS methods to study the atomic-level dynamics of the size evolution of Au_{25} NCs and the seeded growth of Au_{44} from Au_{25} [126]. Overall, all the above studies demonstrate that MS is an indispensable tool in metal NCs research.

3.3.2   TRANSMISSION ELECTRON MICROSCOPY (TEM)

TEM is one of the most powerful tools to characterize NPs, including their size, shape, and crystal structure. But for NCs, it cannot directly image the interfacial ligands and identify the precise core structure. Moreover, melting of tiny NCs under electron beam heating could occur during the imaging process. Nevertheless, high-resolution TEM (HRTEM) is a useful tool and can be performed as a complementary characterization to evaluate the size and dispersity of the obtained NCs in early stages [113, 114].

With the development of aberration-corrected TEM, this technique can be used to carry out single-particle 3D reconstruction analysis to determine the size and structure of Au NCs at atomic resolution. The Kornberg group [127] applied aberration-corrected TEM to image the single Au_{68} NC protected by mcta-mercaptobenzoic acid (m-MBA) at random orientations with low-dose procedures. These 2D images were then combined to determine the 3D atomic

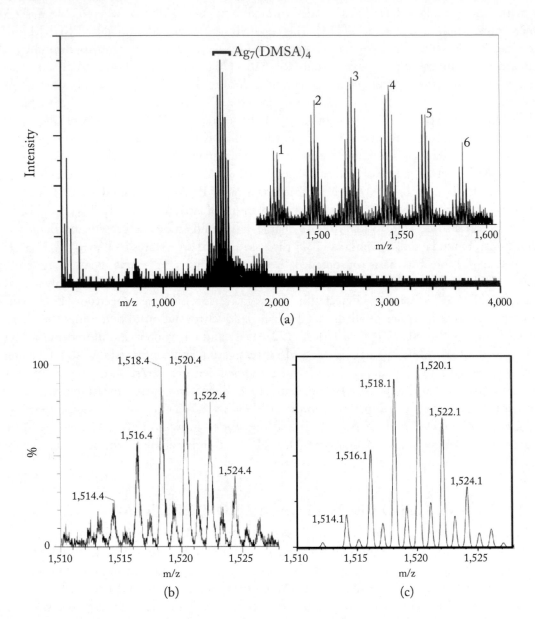

Figure 3.9: (a) ESI-MS spectra of a Ag_7 NC (negative ion mode, inset shows the zoom-in spectrum). (b) and (c) show the experimental and simulated isotopic pattern of $Ag_7L_4 -2H +2Na$, respectively. Reproduced with permission from [114]. Copyright 2009 American Chemical Society.

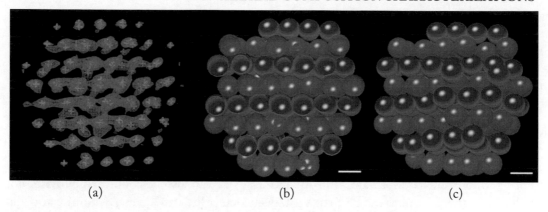

(a)   (b)   (c)

Figure 3.10: (a) Electron density map from 3D reconstruction of presumptive $Au_{144}(m-MBA)_x$ NC, the peaks identified by a peak-search routine, corresponding to locations of gold atoms, indicated by pink crosses. (b) Central section of the structure showing atoms packed with fcc symmetry. (c) Positions of the outermost atoms of the gold core showing deviations from fcc symmetry. Scale bar = 2 Å. Reproduced with permission from [128]. Copyright 2017 American Chemical Society.

structure of the gold NC. The 3D atomic model of the Au_{68} NC reveals that a gold atom at the center is surrounded by a cage-like cuboctahedron of 12 gold atoms. Additional 24 gold atoms form an fcc structure shell around the cuboctahedron, whereas the remaining 31 gold atoms deviate from the fcc packing. Because the surface ligands cannot be measured by this method, 32 sulfur atoms were manually added to the final model based on the positions of the gold atoms and the local stereochemistry. The 3D atomic model of Au_{68} NC determined from this experiment exhibits lower symmetry than that of the Au_{102} NC and is also different from the theoretical prediction of the highly symmetrical atomic structure of a similar NC, $Au_{67}(SR)_{35}$. Nevertheless, the Au_{68} work shows that single-particle 3D reconstruction can be used to determine the atomic structure of metal NCs that are difficult to crystallize [127]. Subsequently, the same group determined the structure of $Au_{144}(m-MBA)_x$ NC [128]. The electron density map displayed 144 well-resolved peaks with spacing of 2.75–3.09 Å, consistent with gold—gold bond lengths of 2.7–3.0 Å (Fig. 3.10a). Most of the 144 gold atoms adopted fcc packing (Fig. 3.10b) and there were also deviations from fcc packing on the surface (Fig. 3.10c), conferring curvature on the NC. Note that such a m-MBA protected fcc Au_{144} NC has an entirely different structure from that of the $Au_{144}(SCH_2Ph)_{60}$ NC, which is revealed to be of the multiple shell-by-shell structure, implying the influence of ligands on NCs structure [128].

Scanning transmission electron microscopy (STEM) is also used for observing the three-dimensional structure of fcc packing metal NCs. Liao and co-workers observed the $6 \times 6 \times 5$ Au atom layers of $Au_{92}(SR)_{44}$ NC in the STEM image [129]. The Dass group characterized the shapes and fcc-strucutures of a series of large plasma gold NCs including $Au_{\sim 500 \pm 10}(SR)_{\sim 120 \pm 3}$ [130], $Au_{940 \pm 20}(SR)_{\sim 160 \pm 4}$ [131], $Au_{\sim 1400}$ [132], and $Au_{\sim 2000}$ [108]. Yang et al. revealed the five-fold twinned structural feature of Ag_{136} and Ag_{374} [133].

3.3.3   X-RAY PHOTOELECTRON SPECTROSCOPY (XPS)

XPS is applied to investigate the elemental composition and oxidation state, and the bonding characteristics in metal NCs. In a typical XPS measurement, monoenergetic soft X-rays are used to irradiate the NC sample and the energies of the ejected electrons are measured. Moreover, the stability of metal NCs can be studied by monitoring the change in the XPS spectra. The presence and location of doped atoms in Au and Ag NCs can also be confirmed by XPS. For example, Wu's group employed XPS to quantitatively detect the Au/Ag/Hg/S atomic ratio and probe the position of Hg atom in $Au_{24-x}HgAg_x(SR)_{18}$ NC [75].

The XPS technique has been extensively used to verify the oxidation states of Au NCs. Negishi et al. studied the Au 4f core-level photoemission spectra of the different-sized $Au_n(SG)_m$ NCs, Au(I) thiolate complex, and Au(0) film. The Au ($4f_{7/2}$) binding energies were found to monotonically shift from 84.0–84.3 eV with the reduction of the core size [112]. Shichibu et al. found that $Au_{18}(SG)_{14}$ turned into insoluble products after reacting with glutathione. The peak position is located between those of $Au_{18}(SG)_{14}$ and sodium Au(I) thiomalate, which suggests that a Au(I):SG complex or polymer is formed during the reaction [134]. Liao et al. found that the Au 4f binding energies of $Au_{44}(2,4-DMBT)_{26}$ are distinctly higher than those of $Au_{44}(TBBT)_{28}$ and even slightly higher than the Au 4f binding energies of the Au-(2,4-DMBT) complex, regardless of the fact that Au-TBBT and Au-(2,4-DMBT) complexes have very closed Au 4f binding energies (2,4-DMBT = 2,4-dimethylbenzenethiol) (Fig. 3.11) [135]. The high Au 4f binding energies of $Au_{44}(2,4-DMBT)_{26}$ indicated the variety of Au charge states in gold NCs. In the electrochemical synthesis of copper NCs, the binding energy peaks from both Cu(0) and Cu(II) can be observed in the XPS measurements. After the Ar^+ ion sputtering treatment, the XPS signals corresponding to Cu(II) disappeared completely and only Cu(0) and a small amount of Cu(I) remains in the NCs [136].

3.3.4   THERMOGRAVIMETRIC ANALYSIS (TGA)

TGA is commonly used in the initial stages of characterization of noble metal NCs. Under thermal conditions, NCs lose surface ligands and transform to pure metal. TGA, when it is used in combination with some conventional chemical analyses, can give an initial indication of the composition of metal NCs. The Murray group successfully used TGA to calculate the percentage of organic ligands (e.g., thiolate monolayers) in monolayer-protected NCs [137]. The Wu group used TGA to support the component identification of monometallic and doped Au NCs [72, 135]. Levi-Kalisman et al. studied the ligand coverage of $Au_{102}(p-MBA)_{44}$ NCs on the basis of TGA results [138]. The weight loss (%) vs. temperature (°C) curve showed a major decrease in mass between 260 and 360°C. While the major ($\sim 23\%$) and total loss ($\sim 30\%$)

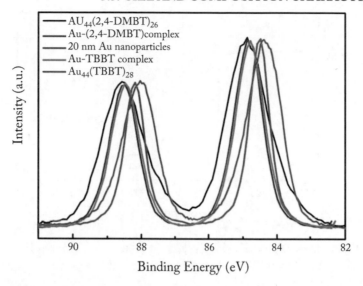

Figure 3.11: Au 4f XPS spectra of $Au_{44}(2,4-DMBT)_{26}$, $Au-(2,4-DMBT)$ complex, ~ 20 nm Au NPs, Au-TBBT complex, and $Au_{44}(TBBT)_{28}$. Reproduced with permission from [135]. Copyright 2016 American Chemical Society.

exhibited some discrepancy from the calculated result for $Au_{102}(p-MBA)_{44}$ NCs, which could be ascribed to the contamination by particles with other compositions, or inaccuracy of the procedure. So, the sample for TGA measurement should be highly dried and purified.

The stability of metal NCs can be also evaluated by TGA. In the work of Jin's group, TGA shows that three $Au_n(SR)_m$ NCs [including $Au_{25}(SR)_{18}$, $Au_{38}(SR)_{24}$, and $Au_{144}(SR)_{60}$] start to lose ligands at ~ 200°C irrespective of size, and the ligand loss typically completes at ~ 250°C. The ligand-loss temperature was unaffected by the atmosphere (e.g., N_2, air, O_2, and H_2) [139]. Thus, they chose 150°C as the temperature for annealing the wet-deposited $Au_{25}(SR)_{18}$/oxide catalysts and no weight loss was found, indicating that the gold NCs remain stable during the annealing process [140]. The Negishi group studied the stability of solid $[Au_{25}(SC_8H_{17})_{18}]^-$ and $[Au_{25}(SeC_8H_{17})_{18}]^-$ against thermal dissociation. TGA results show that the temperature at which the ligands start to evaporate are 165°C for $[Au_{25}(SC_8H_{17})_{18}]^-$ and 136°C for $[Au_{25}(SeC_8H_{17})_{18}]^-$. Considering that the Se–C bond energy (590.4 kJ/mol) is lower than the S–C bond energy (713.3 kJ/mol), this result indicates that the ligand change from thiolate to selenolate reduces the stability of NCs against intramolecular dissociation (Fig. 3.12) [141].

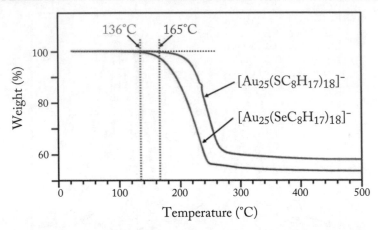

Figure 3.12: TGA curves of $[Au_{25}(SC_8H_{17})_{18}]^-$ and $[Au_{25}(SeC_8H_{17})_{18}]^-$. Reproduced with permission from [141]. Copyright 2012 American Chemical Society.

3.4   OPTICAL PROPERTY CHARACTERIZATIONS

3.4.1   UV-Vis ABSORPTION SPECTROSCOPY

It has been well known that the UV-Vis absorption of coinage metal NPs (Au, Ag, and Cu) are dominated by surface plasmon resonance (SPR) peaks at around 520, 400, and 600 nm for Au, Ag, and Cu NPs, respectively [142, 143]. Significantly different from large NPs, the UV-Vis absorption of metal NCs exhibit molecular-like optical transitions with absorbance bands due to the quantum confinement effects [14]. Based on the structure-sensitive optical characteristics of NCs, UV-Vis absorption spectroscopy has been used as a powerful and convenient diagnostic tool to study the optical absorption and electronic structure of metal NCs.

The Jin and Murray groups synthesized $Au_{25}(SR)_{18}$ NCs with distinct absorption bands at 1.82, 2.75, and 3.10 eV irrespective of the −R groups (Fig. 3.13) [22, 102, 144]. The multiple molecular-like transitions in the absorption spectrum are ascribed to the strong quantum size effects of $Au_{25}(SR)_{18}$ NCs, which are different from the surface plasmon resonance at ∼ 2.4 eV for large gold NPs. Based on the crystal structure, Aikens et al. carried out time-dependent density functional theory (TD-DFT) calculations and reproduced the experimental spectrum. Both the sp and d bands were found to be quantized. The experimental peak at 1.82 eV (peak a in Fig. 3.13) was revealed to be the HOMO-to-LUMO transition, which is essentially an interband ($sp \leftarrow sp$) transition [22]. As for the peak of 3.10 eV (peak c in Fig. 3.13), the transition mode is an intraband transition ($sp \leftarrow d$), arising from the other occupied HOMO to the n orbitals (these orbitals belong to the d band). The peak at 2.75 eV (peak b in Fig. 3.13) is composed of the mixed types of transitions (containing the interband $sp \leftarrow sp$ transition and the intraband $sp \leftarrow d$ transition). Taken together, the optical absorption properties of Au_{25} NCs

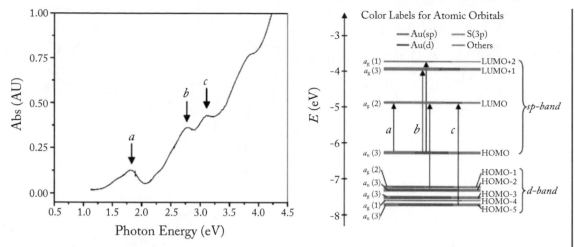

Figure 3.13: Peak assignments of the absorption spectra and Kohn–Sham orbital level diagram of the $Au_{25}(SR)_{18}$ NC. Reproduced with permission from [22]. Copyright 2008 American Chemical Society.

apparently root in strong quantum confinement of electrons in the cluster. This NC serves as a good example to illustrate the quantum size effect on the optical properties of gold NCs.

Ag and Cu NCs also show distinct absorption peaks in the UV-vis region, and their optical absorption can be readily characterized by this instrument. Bakr et al. reported the synthesis of Ag NCs through the reduction of a silver salt in the presence of the capping ligand 4-fluorothiophenol. The UV-Vis absorption spectrum of the prepared NCs exhibits six pronounced and two weak peaks which are markedly different from those previously reported for either silver NCs or silver NPs [145]. Chen's group synthesized stable Cu_n ($n \leq 8$) NCs with three well-defined absorption bands at 285, 364, and 443 nm, which are clearly different from the characteristic SPR band of large Cu NPs around 560–600 nm [115]. Similar to the gold NCs, the multiband absorption of Cu NCs are also from the interband or intraband electronic transitions due to the discrete energy levels.

The shape of UV-vis absorption spectrum can be influenced by temperature, the sample purity, and the ligand type. Ramakrishna and co-workers monitored the temperature-dependent optical absorption spectra of $Au_{25}(SR)_{18}$ NC from 303 K down to 90 K. The absorption peaks of $Au_{25}(SR)_{18}$ became sharper and showed significant enhancement of intensity accompanying the decrease in temperature (Fig. 3.14) [146]. Weissker et al. revealed how sample cooling brings forth a clear and well-resolved multiple-band structure in Au_{144} NCs [147]. Comparison of the UV-vis absorption spectra of $Au_{144}(SR)_{60}$ at room temperature and 77 K shows clearly the development of the peak shape on temperature, indicating the thermal broadening obscures the intrinsic fine structure of the spectrum [147]. Yan et al. found that the absorption peaks of the

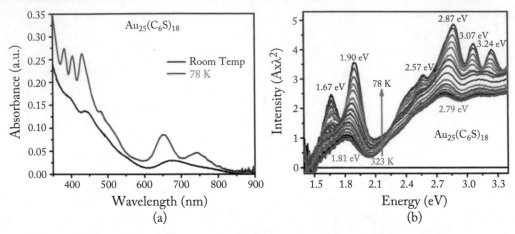

Figure 3.14: (a, b) The temperature-dependent optical absorptions of $Au_{25}(SR)_{18}$ NC. Reproduced with permission from [146]. Copyright 2011 American Chemical Society.

redissolved crystals of $Au_{144}(SCH_2Ph)_{60}$ were more distinct than the solution of the amorphous solids of $Au_{144}(SCH_2Ph)_{60}$, probably because of the small-molecule impurities not identified by MS [31]. Moreover, a red shift of the absorption peaks of $Au_{144}(SCH_2Ph)_{60}$ compared with the peaks of $Au_{144}(SC_2H_4Ph)_{60}$ was observed, attributing to the ligand effect [31]. Sementa et al. investigated ligand-enhanced optical absorption of $Au_{25}(SR)_{18}$ NC according to the TD-DFT calculation [148]. By tailoring the steric hindrance and electronic conjugating features of ligands, notable enhancements in optical absorption and effective electron delocalization were simultaneously achieved [148].

Due to the convenient operation in experiment, UV-vis absorption spectroscopy can be used to identify the metallic or molecular-like states of NCs, as well as monitor the reaction process and provide information about the reaction mechanism. For example, Negishi et al [149], compared the optical absorption spectra of a series of $Au_n(SC_{12})_m$ in the n range from 38 to ~ 520. The small NCs ($n < 144$) have multiple absorption peaks and discrete electronic structures. On the other hand, the optical absorption spectra of large NCs ($n > 187$) exhibit a distinct single peak in the 520–540 nm range assignable to SPR. Therefore, they concluded that the transition from molecular-like state to metallic state occur between $Au_{144}(SC_{12})_{60}$ and $Au_{187}(SC_{12})_{68}$ [149]. Wu et al. found, in the synthesis of Au_{25} NCs, that the initial polydisperse product eventually converted to the truly monodisperse Au_{25} NC when the reaction time was extended, evidenced by the evolution of the UV-vis spectrum (Fig. 3.15) [124]. Such an interesting phenomenon was termed as "size-focusing."

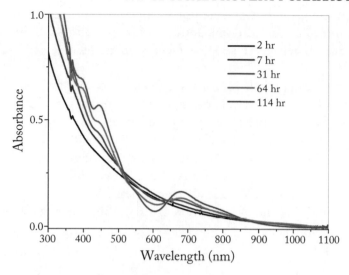

Figure 3.15: Evolution of the UV-vis spectra of the crude product with aging time. The spectra are vertically shifted for the ease of comparison. Reproduced with permission from [124]. Copyright 2009 Royal Society of Chemistry.

3.4.2   PHOTOLUMINESCENCE (PL) SPECTROSCOPY

If the metal NCs are excited from the ground state, they will release extra energy before returning to the ground state, which generates photoluminescence (PL). The PL of metal NCs has attracted wide research interest because of their promising applications in cell labeling, biosensing, phototherapy, and so on. PL spectroscopies include steady-state, temperature-dependent, and time-resolved PL spectroscopies. Among such techniques, the steady-state PL analysis is mostly used and the related research results have been summarized in several reviews [15, 18, 150, 151]. Here we briefly introduce the application of the latter two methods.

Temperature-dependent PL can provide unique physical insight into the nature of photoluminescent metal NCs. Yu et al. observed temperature-dependent PL in histidine-protected Au_{10} NCs in the temperature range from 77–300 K [152]. The intensity of luminescence in Au_{10} decreased with increasing temperature, which was attributed to the thermally activated nonradiative trapping and the activation energy was determined to be 103 meV. Based on the observation and comparison with the Au NPs, they suggested that, other than electron–phonon scattering, electron-electron interactions and surface/defect/impurity scattering play a dominant role in Au_{10} [152]. Wen et al. investigated the fluorescence origin of the Au_{25}@BSA (BSA = bull serum albumin) [153]. They elucidated that the fluorescence of Au_{25}@BSA consists of two bands (Fig. 3.16a). The band I exclusively originates from the Au_{13} icosahedral core and exhibits similarity to semiconductors. Its temperature-dependent energy gap and bandwidth can be well fitted using the equations that were usually used in semiconductor QDs, as shown in Fig. 3.16b-c. In contrast, the band II dominantly arises from the S-Au-S-Au-S semirings. The energy gap and bandwidth of band II exhibit evidently different temperature effects from those in band I [153].

Time-resolved PL can provide information about the carrier dynamics, which is critical for the understanding of various physical processes and PL mechanism. Time-correlated single photon counting (TCSPC) has been widely applied for the investigation of PL lifetimes in metal NCs for timescales of ns-μs. In smaller Au ($Au_8 - Au_{13}$) and Ag ($< Ag_9$) NCs, lifetimes of few ns were usually observed and are attributed to the recombination of electron-hole pairs. In contrast, the larger Au and Ag NCs, such as Au_{25} and Au_{38}, exhibit evidently slower lifetimes in the μs range [15]. For instance, Wen et al. systematically investigated the carrier dynamics in Au_{25}@BSA and found that the PL evolution of Au_{25}@BSA includes a fast ns component followed by a slow μs component [154].

3.4.3   ULTRAFAST ELECTRON DYNAMICS

Metal NCs display molecule-like physicochemical properties that are different than those of larger NPs. The ultrafast electron relaxation processes of the NCs show some similarities as in small molecules, but not totally the same. In recent years, research on the ultrafast electron dynamics, including the relaxation time scales, radiative emission and electron-phonon coupling, have been widely studied because of their importance in the understanding of the electronic structure and practical applications of NCs. Several techniques have been employed to understand the correlations of the ultrafast electron dynamics and structures of NCs, such as femtosecond time-resolved fluorescence up-conversion, transient absorption, and other methods [155–157].

Qian et al. reported the effects of charge states on the ultrafast relaxation dynamics of $Au_{25}(SR)_{18}^q$ with different charge states. For $Au_{25}(SR)_{18}$ with $q = 0$ and -1 charge states, photo excitation occurred in two non-degenerate states near the HOMO–LUMO gap originating from the core orbitals. However, the lifetimes of the core excitations of these two NCs showed obvious differences. The decay rate of anionic $[Au_{25}(SR)_{18}]^-$ was > 1000 times slower than that of the NC with neutral state (Fig. 3.17) [158]. They proposed that the differences mainly originated from three aspects: (1) different absolute energies of the transitions; (2) enhanced symmetry of $[Au_{25}(SR)_{18}]^0$; and (3) the presence of a counterion in the $[Au_{25}(SR)_{18}]^-$ NC. Thomas and Knappenberger also discussed the change of relaxation dynamics of Au_{25} NCs with different charge states. They further found that the NIR excited state absorption measurements were helpful in describing the relaxation processes of metal NCs [159]. Recently, Zhou et al. reported a three-orders-of-magnitude variation of carrier lifetimes in exotic crystalline phases of gold NCs, including hcp Au_{30} and bcc Au_{38} NCs protected by the same type of capping ligand. By comparing the transient absorption spectra, they found that the bcc Au_{38} NC had an

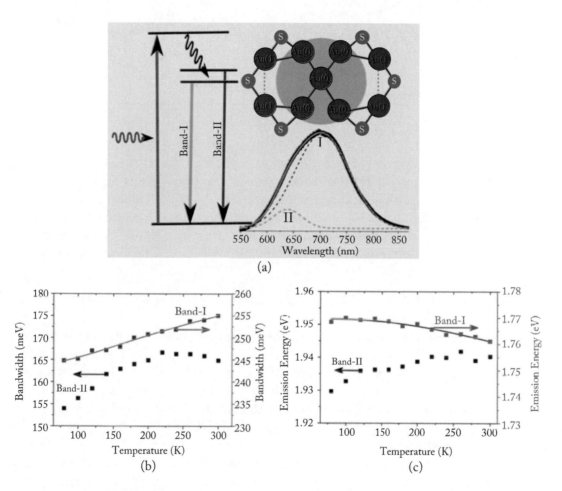

Figure 3.16: (a) A schematic diagram shows the red fluorescence of Au$_{25}$ NCs originates from two transitions that is correlated to the structure. (b) With increasing temperature, the energy gap of band I exhibits a red shift of 11 meV. In contrast, band II displays a blue shift of 10 meV. (c) The bandwidth exhibits temperature-dependent variation. Reproduced with permission from [153]. Copyright 2012 American Chemical Society.

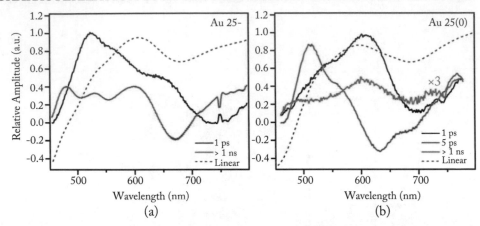

Figure 3.17: (A) Global analysis of broad-band transient absorption spectra for $[Au_{25}(SR)_{18}]^-$ excited at 390 nm. Two main kinetic components are found (a ~ 1 ps lifetime (black) and a long lifetime (> 1 ns, red)). The linear absorption spectrum (gray dotted line) is plotted on an arbitrary reverse axis for comparison. (B) Three-component global fit for $[Au_{25}(SR)_{18}]^0$ excited at 420 nm. Reproduced with permission from [158]. Copyright 2010 American Chemical Society.

exceptionally long carrier lifetime (4.7 μs) comparable to that of bulk silicon, whereas the hcp Au_{30} NC had a very short lifetime (1 ns) [160].

Furthermore, electron dynamics can also be used to differentiate molecular and metallic behaviors by monitoring the pump-power dependence. Unlike the strong power-dependent electron–phonon coupling of the Au NPs, laser-power-independent electron dynamics was observed on the Au NCs [17]. For example, the Whetten group [161] investigated the femtosecond transient absorption of Au NCs composed of 28 gold atoms and GSH ligands (later corrected to $Au_{25}(SG)_{18}$). The excited-state relaxation of this NC exhibited biexponential decay with a sub-picosecond and a longer nanosecond decay time, independent of the laser pump power. In contrast, the excited-state relaxations of larger NPs showed a much shorter decay. Specifically, the transient absorptions of the $Au_{25}(SG)_{18}$ NCs were composed of excited-state absorption, which is different from the strong bleaching due to SPR in the NPs [161]. With the same principle, Higaki et al. revealed a sharp transition from nonmetallic Au_{246} to metallic Au_{279} with femtosecond transient absorption spectroscopy [162].

3.4.4   CHARACTERIZATION OF OPTICAL ACTIVITY

In recent years, chirality has become an intensive research topic for nanoscale materials. Chirality is a ubiquitous phenomenon in nature. Generally, when an object and its mirror image cannot be superimposed, the object is chiral and gives rise to optical activity (e.g., circular dichroism and Faraday rotation).

For metal NCs, the optical activity is typically probed with circular dichroism (CD) spectroscopy and, sometimes, with vibrational circular dichroism (VCD), along with NMR spectroscopy. The basic principle of CD is the differential absorption of left- and right-circularly polarized light by a chiral sample, whereas VCD is sensitive to the absolute configuration as well as the conformation of a chiral molecule in solution. Achiral molecules or racemic mixtures do not show any VCD activity due to the fact that two enantiomeric forms of the chiral conformation are equally abundant without a chiral environment [163, 164].

In early studies before the successful determination of structures of thiolated gold NCs, it was speculated that chirality might originate from the inner Au kernel [166]. Since the reports of crystal structures, the research on chirality of metal NCs has led to significant achievements [163, 166]. Jin et al. summarized the four origins of the chirality of thiolated gold NCs: (i) chiral arrangement of surface units; (ii) inherent chirality of the cluster kernel; (iii) chiral arrangement of carbon tails; and (iv) chiral thiolate ligands. Note that the last origin can induce chirality to the achiral gold NCs [1, 168]. For example, the Au_{25} framework has a symmetry plane (D_{2h} point group) and is achiral, but by introducing different chiral ligands, rich CD spectral signals of the Au_{25} NCs were reported (Fig. 3.18) [169]. In contrast, the chirality of the cluster kernel can also affect the achiral ligand. Buürgi and co-workers found that the chirality of the Au_{38} core can transfer to the achiral $-SC_2H_4Ph$ ligand, resulting in a chiral conformation of the ligand, evidenced by the observation of strong VCD signals of the surface ligands [170].

The enantiomers of chiral NCs can be separated through chiral HPLC and then examined by CD spectroscopy. Bürgi and co-workers succeeded in the first HPLC isolation of the racemic $Au_{38}(SR)_{24}$ NC [171]. Enantiomeric separation was also reported for $Au_{28}(SR)_{20}$ [43] and $Au_{40}(SR)_{24}$ [172].

3.5 CHARACTERIZATION OF ELECTROCHEMICAL PROPERTIES

The electrochemical characterization of metal NCs is crucial for understanding the redox behavior and determining the energies of the HOMO and LUMO. The HOMO–LUMO gap (i.e., the difference between the first oxidation and reduction potentials after subtraction of the charging energy), the energy levels and the redox potential of the NCs are closely related to the chemical reactivity and electrocatalysis applications of NCs [173, 174].

Three methods are most frequently used for investigating the electrochemical properties as well as the HOMO–LUMO gaps of NCs, including the cyclic voltammetry (CV), differential pulse voltammetry (DPV), and square-wave voltammetry (SWV).

In 2004, the Murray group [175] reported the DPV and CV of the $Au_{25}(SC_2H_4Ph)_{18}$ NC (note: erroneously formulated as $Au_{38}(SC_2H_4Ph)_{24}$ due to the lack of MS characterization at that time, but later corrected). From the DPV results of the anionic $Au_{25}(SR)_{18}$ NC in the CH_2Cl_2 at 25°C, the potential spacing between the first oxidation and the first reduction state is calculated to be 1.6 V, representing the HOMO–LUMO gap (without subtracting the charging

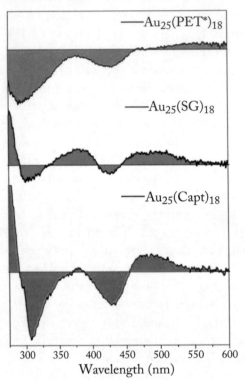

Figure 3.18: Effect of the chiral ligand type on the CD spectra of $Au_{25}(SR^*)_{18}$ (where the asterisk indicates chiral carbon tails). Reproduced with permission from [169]. Copyright 2012 Royal Society of Chemistry.

energy, typically $\sim$ 0.2–0.3 eV) [175]. Lee and co-workers characterized the electrochemical properties of both organic- and water-soluble Au_{25} NCs. The water-soluble Au_{25} NC protected by (3-mercaptopropyl) sulfonate is capable of transferring into the organic phase by ion-pairing the terminal sulfonate groups of the surface ligands with hydrophobic counterions. The phase-transferred Au_{25} (denoted as PT-Au_{25}) NC was stable in the organic solvent. By comparing the DPV results of Au_{25} and PT-Au_{25}, the potential of the oxidation peak was maintained; however, the reduction peak was red-shifted from -2.15 V of water-soluble Au_{25} to -1.85 V of PT-Au_{25}. In addition, the polarity of the solvent had a significant impact on the SWV spectra of PT-Au_{25}; in contrast, no changes were found from the SWV of the water-soluble Au_{25} NC [176].

Liao et al. conducted the DPV measurements on three $Au_{25}(SR)_{18}$ species protected by various ligands and $Au_{24}Hg(PET)_{18}$ (Fig. 3.19) [72]. They found that the first oxida-tion/reduction potential values in DPV, together with the HOMO/ LUMO energy values for

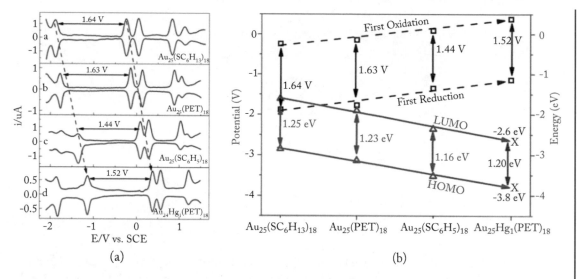

Figure 3.19: (a) DPV spectra and (b) summary of the first oxidation/reduction potential values and the HOMO/LUMO energies of $Au_{25}(SC_6H_{13})_{18}$, $Au_{25}(PET)_{18}$, $Au_{25}(SC_6H_5)_{18}$, and $Au_{24}Hg_1(PET)_{18}$ at 0.01 V/s in degassed CH_2Cl_2 containing 0.1 M Bu_4NPF_6 with 1-mm diameter Pt working, SCE reference, and carbon rod counter electrodes. Reproduced with permission from [72]. Copyright 2015 American Chemical Society.

the three Au_{25} species are linearly related with the polarity of the ligand. Thus, the HOMO and LUMO energies of $Au_{24}Hg_1(PET)_{18}$ are readily deduced based on the linear relation.

Besides, the electrochemical methods can also be used to investigate the mechanism of unprecedented reactions of NCs or NPs, especially the anti-galvanic reaction (AGR)—which means that the reduction of metal ions by less reactive metals [177]. Employing the CV test, Wu et al. demonstrated that the oxidation potential of surfactant- and ligand-free gold NPs (-0.76 V vs. Ag^+/Ag) is obviously lower than the reduction potential of Ag^+ (-0.21 V vs. Ag^+/Ag), which leads to the oxidation of gold NPs and the reduction of Ag^+ [178]. Jin et al. also revealed the similar oxidation potential decrease of $\sim$ 2-nm gold NCs protected by 11-mercaptoundecanoic acid (11-MUA) [179]. Recently, Zhang et al. reported a quasi-AGR between $Au_{26}Cd_4(SR)_{22}$ NC and Au-SR complex. The redox reactions are validated by CV measurements [180].

3.6 CHARACTERIZATION OF MAGNETIC PROPERTIES

Single gold atoms are paramagnetic due to the 6s [1] unpaired electron, whereas bulk gold is diamagnetic due to the electronic band formation. The magnetism evolution from single gold atoms to metallic NPs is of fundamental interest and importance in the understanding of the

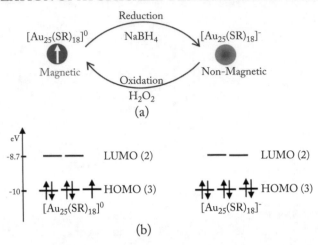

Figure 3.20: (a) Reversible conversion between neutral and anionic $Au_{25}(SR)_{18}$ NCs. (b) DFT-calculated Kohn–Sham orbital energy level diagrams for neutral and anionic NCs, respectively. Reproduced with permission from [182]. Copyright 2009 American Chemical Society.

magnetism of gold [181]. As the bridge linking up atoms and larger NPs, Au NCs are considered to be the ideal system for studying the magnetic origin and evolution.

The main characterization method of the magnetic properties of the NC is electron paramagnetic resonance (EPR), which can detect the unpaired spin in the particles. In 2009, Zhu et al. [182] reported that the EPR spectra of microcrystalline and frozen solutions of $[Au_{25}(SR)_{18}]^0$ NC showed an $S = 1/2$ signal, which suggests that each NC is paramagnetic and has one unpaired spin. The EPR signal disappeared upon the reduction of the NC to its negative charge state but could be switched back when the negatively charged NC was oxidized to the neutral state. With the available crystal structure of $Au_{25}(SR)_{18}$, Zhu et al. achieved a correlation between the structure and magnetic properties. DFT calculations indicated that the unpaired spin is in the highest occupied Kohn–Sham orbital in the neutral Au_{25} NC, which is mainly the origin of the observed magnetism (Fig. 3.20) [182]. The Wu group investigated the magnetism of mono-doped MAu_{24} NCs using EPR. The doped atoms (M = Hg, Pd, Pt) can affect the electron configuration of $[MAu_{24}(SR)_{18}]^0$ and transform the paramagnetism to diamagnetism [72, 89].

Superconducting quantum interference device (SQUID) magnetometer is also used to analyze the magnetism of NCs. For example, the Jin group reported the one-pot synthesis of water soluble $Au_{25}(SG)_{18}^-$, 2-nm and 4-nm NPs. Both the 2-nm and 4-nm Au NPs exhibit paramagnetism, while the $Au_{25}(SG)_{18}^-$ is diamagnetic [183]. Krishna et al. determined the paramagnetism of $[Au_{25}(SC_2H_4Ph)_{18}]^0$, the diamagnetism of $[Au_{25}(PPh_3)_{10}(SC_{12}H_{25})Cl_2]^{2+}$ and $Au_{38}(SC_{12}H_{25})_{24}^0$, and the ferromagnetism of $Au_{55}(PPh_3)_{12}Cl_6$ [184].

Besides, some other magnetic characterization methods were also applied. Negishi et al. investigated a series of atomically precise NCs with a general formula $Au_n(SG)_m$, where $(n, m) = (10, 10), (15, 13), (22, 16), (22, 17), (25, 18), (29, 20)$, and $(39, 24)$, via X-ray magnetic circular dichroism (XMCD) analysis. The magnetic moment was found to increase with an increase in the NC core size. The hole created by Au–S bonding is more responsible than the quantum size effect of gold NCs for the spin polarization phenomenon [185]. Maran and co-workers assessed the interactions between the metallic kernel and the capping ligands of paramagnetic $Au_{25}(SR)_{18}^0$ NC by studying the interactions of unpaired electrons with pulse electron nuclear double resonance (ENDOR) spectroscopy [186].

3.7   CONCLUSION

In this chapter, we summarized the common characterization methods for atomically precise metal NCs, including the structure determination, size and composition analyses, and studies on the optical, electrochemical, and magnetic properties. Various analytical methods have been developed for the full characterization of metal NCs, which has not only deepened the understanding of such new materials, but also promoted the practical applications of NCs in the biomedical, energy, environmental, and catalysis fields (see Chapter 4). One of the most important methods is certainly SCXRD, by which a large number of crystal structures of mono- and multi-metallic NCs have been solved. Such atomic-level structures provide critical information for correlating the structure-function relationships.

Some challenging issues still remain, which limit the further research and applications of NCs, and therefore require future efforts from the community. For example, the structural identification of large plasmonic NCs ($> Au_{279}$) is still difficult, which hinders research on the evolution of metallic state. The origins of PL in NCs are still not fully understood. No ferromagnetism has been found in well-defined thiolated Au NCs, although it was reported in imprecise NPs. The solutions to these problems will depend on the further development of characterization methods as well as the precision synthesis and theoretical investigation. We believe that the future advances will shed light on the above issues.

3.8   REFERENCES

[1] Jin, R., Zeng, C., Zhou, M., and Chen, Y. Atomically precise colloidal metal nanoclusters and nanoparticles: Fundamentals and opportunities. *Chem. Rev.*, 116:10346–10413, 2016. DOI: 10.1021/acs.chemrev.5b00703. 31, 32, 37, 57, 59

[2] Chakraborty, I. and Pradeep, T. Atomically precise clusters of noble metals: Emerging link between atoms and nanoparticles. *Chem. Rev.*, 117:8208–8271, 2017. DOI: 10.1021/acs.chemrev.6b00769. 31, 37

[3] Jin, R., Qian, H., Wu, Z., Zhu, Y., Zhu, M., Mohanty, A., and Garg, N. Size focusing: A methodology for synthesizing atomically precise gold nanoclusters. *J. Phys. Chem. Lett.*, 1:2903–2910, 2010. DOI: 10.1021/jz100944k. 31

[4] Zeng, C., Chen, Y., Das, A., and Jin, R. Transformation chemistry of gold nanoclusters: From one stable size to another. *J. Phys. Chem. Lett.*, 6:2976–2986, 2015. DOI: 10.1021/acs.jpclett.5b01150. 31

[5] Gan, Z., Xia, N., and Wu, Z. Discovery, mechanism, and application of antigalvanic reaction. *Acc. Chem. Res.*, 51:2774–2783, 2018. DOI: 10.1021/acs.accounts.8b00374. 31

[6] Zheng, K., Yuan, X., Goswami, N., Zhang, Q., and Xie, J. Recent advances in the synthesis, characterization, and biomedical applications of ultrasmall thiolated silver nanoclusters. *RSC Adv.*, 4:60581–60596, 2014. DOI: 10.1039/c4ra12054j. 31

[7] Udayabhaskararao, T. and Pradeep, T. New protocols for the synthesis of stable Ag and Au nanocluster molecules. *J. Phys. Chem. Lett.*, 4:1553–1564, 2013. DOI: 10.1021/jz400332g. 31

[8] Sharma, S., Chakrahari, K. K., Saillard, J.-Y., and Liu, C. W. Structurally precise dichalcogenolate-protected copper and silver superatomic nanoclusters and their alloys. *Acc. Chem. Res.*, 51:2475–2483, 2018. DOI: 10.1021/acs.accounts.8b00349. 31, 37

[9] Lu, Y. and Chen, W. Sub-nanometre sized metal clusters: From synthetic challenges to the unique property discoveries. *Chem. Soc. Rev.*, 41:3594–3623, 2012. DOI: 10.1039/c2cs15325d. 31

[10] Liu, X. and Astruc, D. Atomically precise copper nanoclusters and their applications. *Coord. Chem. Rev.*, 359:112–126, 2018. DOI: 10.1016/j.ccr.2018.01.001. 31

[11] Lu, Y. and Chen, W. Application of mass spectrometry in the synthesis and characterization of metal nanoclusters. *Anal. Chem.*, 87:10659–10667, 2015. DOI: 10.1021/acs.analchem.5b00848. 31, 42

[12] Harkness, K. M., Cliffel, D. E., and McLean, J. A. Characterization of thiolate-protected gold nanoparticles by mass spectrometry. *Analyst*, 135:868–874, 2010. DOI: 10.1039/b922291j. 31, 42

[13] Jadzinsky, P. D., Calero, G., Ackerson, C. J., Bushnell, D. A., and Kornberg, R. D. Structure of a thiol monolayer-protected gold nanoparticle at 1.1 Å Resolution. *Science*, 318:430–433, 2007. DOI: 10.1126/science.1148624. 31, 32, 33

[14] Jin, R. Atomically precise metal nanoclusters: Stable sizes and optical properties. *Nanoscale*, 7:1549–1565, 2015. DOI: 10.1039/c4nr05794e. 31, 50

[15] Yu, P., Wen, X., Toh, Y.-R., Ma, X., and Tang, J. Fluorescent metallic nanoclusters: Electron dynamics, structure, and applications. *Part. Syst. Charact.*, 32:142–163, 2015. DOI: 10.1002/ppsc.201400040. 31, 53, 54

[16] Du, X. and Jin, R. Atomically precise metal nanoclusters for catalysis. *ACS Nano*, 13:7383–7387, 2019. DOI: 10.1021/acsnano.9b04533. 31

[17] Zhou, M., Zeng, C., Chen, Y., Zhao, S., Sfeir, M. Y., Zhu, M., and Jin, R. Evolution from the plasmon to exciton state in ligand-protected atomically precise gold nanoparticles. *Nat. Commun.*, 7:13240, 2016. DOI: 10.1038/ncomms13240. 31, 56

[18] Kang, X. and Zhu, M. Tailoring the photoluminescence of atomically precise nanoclusters. *Chem. Soc. Rev.*, 48:2422–2457, 2019. DOI: 10.1039/c8cs00800k. 31, 53

[19] Du, Y., Sheng, H., Astruc, D., and Zhu, M. Atomically precise noble metal nanoclusters as efficient catalysts: A bridge between structure and properties. *Chem. Rev.*, 120:526–622, 2020. DOI: 10.1021/acs.chemrev.8b00726. 31

[20] Xia, Y., Xia, X., and Peng, H.-C. Shape-controlled synthesis of colloidal metal nanocrystals: Thermodynamic versus kinetic products. *J. Am. Chem. Soc.*, 137:7947–7966, 2015. DOI: 10.1021/jacs.5b04641. 32

[21] Antonello, S., Dainese, T., Pan, F., Rissanen, K., and Maran, F. Electrocrystallization of monolayer-protected gold clusters: Opening the door to quality, quantity, and new structures. *J. Am. Chem. Soc.*, 139:4168–4174, 2017. DOI: 10.1021/jacs.7b00568. 32

[22] Zhu, M., Aikens, C. M., Hollander, F. J., Schatz, G. C., and Jin, R. Correlating the crystal structure of a thiol-protected Au_{25} cluster and optical properties. *J. Am. Chem. Soc.*, 130:5883–5885, 2008. DOI: 10.1021/ja801173r. 33, 50, 51

[23] Heaven, M. W., Dass, A., White, P. S., Holt, K. M., and Murray, R. W. Crystal structure of the gold nanoparticle $[N(C_8H_{17})_4][Au_{25}(SCH_2CH_2Ph)_{18}]$. *J. Am. Chem. Soc.*, 130:3754–3755, 2008. DOI: 10.1021/ja800561b. 33

[24] Qian, H., Eckenhoff, W. T., Zhu, Y., Pintauer, T., and Jin, R. Total structure determination of thiolate-protected Au_{38} nanoparticles. *J. Am. Chem. Soc.*, 132:8280–8281, 2010. DOI: 10.1021/ja103592z. 33

[25] Alvarez, M. M., Khoury, J. T., Schaaff, T. G., Shafigullin, M. N., Vezmar, I., and Whetten, R. L. Optical absorption spectra of nanocrystal gold molecules. *J. Phys. Chem. B*, 101:3706–3712, 1997. DOI: 10.1021/jp962922n. 33, 34, 42

[26] Chaki, N. K., Negishi, Y., Tsunoyama, H., Shichibu, Y., and Tsukuda, T. Ubiquitous 8 and 29 kDa gold: Alkanethiolate cluster compounds: Mass-spectrometric determination

of molecular formulas and structural implications. *J. Am. Chem. Soc.*, 130:8608–8610, 2008. DOI: 10.1021/ja8005379.

[27] Qian, H. and Jin, R. Controlling nanoparticles with atomic precision: The case of $Au_{144}(SCH_2CH_2Ph)_{60}$. *Nano Lett.*, 9:4083–4087, 2009. DOI: 10.1021/nl902300y.

[28] Koivisto, J., Salorinne, K., Mustalahti, S., Lahtinen, T., Malola, S., Häkkinen, H., and Pettersson, M. Vibrational perturbations and ligand—layer coupling in a single crystal of $Au_{144}(SC_2H_4Ph)_{60}$ nanocluster. *J. Phys. Chem. Lett.*, 5:387–392, 2014. DOI: 10.1021/jz4026003. 34

[29] Ackerson, C. J., Jadzinsky, P. D., Sexton, J. Z., Bushnell, D. A., and Kornberg, R. D. Synthesis and bioconjugation of 2 and 3 nm-diameter gold nanoparticles. *Bioconjugate Chem.*, 21:214–218, 2010. DOI: 10.1021/bc900135d.

[30] Zeng, C., Chen, Y., Kirschbaum, K., Lambright, K. J., and Jin, R. Emergence of hierarchical structural complexities in nanoparticles and their assembly. *Science*, 354:1580–1584, 2016. DOI: 10.1126/science.aak9750. 34

[31] Yan, N., Xia, N., Liao, L., Zhu, M., Jin, F., Jin, R., and Wu, Z. Unraveling the long-pursued Au_{144} structure by x-ray crystallography. *Sci. Adv.*, 4:eaat7259, 2018. DOI: 10.1126/sciadv.aat7259. 33, 34, 52

[32] Chen, S., Wang, S., Zhong, J., Song, Y., Zhang, J., Sheng, H., Pei, Y., and Zhu, M. The structure and optical properties of the $[Au_{18}(SR)_{14}]$ nanocluster. *Angew. Chem. Int. Ed.*, 54:3145–3149, 2015. DOI: 10.1002/anie.201410295. 34

[33] Das, A., Liu, C., Byun, H. Y., Nobusada, K., Zhao, S., Rosi, N., and Jin, R. Structure determination of $[Au_{18}(SR)_{14}]$. *Angew. Chem. Int. Ed.*, 54:3140–3144, 2015. DOI: 10.1002/anie.201410161. 34

[34] Higaki, T., Liu, C., Zeng, C., Jin, R., Chen, Y., Rosi, N. L., and Jin, R. Controlling the atomic structure of Au_{30} nanoclusters by a ligand-based strategy. *Angew. Chem. Int. Ed.*, 55:6694–6697, 2016. DOI: 10.1002/anie.201601947. 34

[35] Zeng, C., Liu, C., Chen, Y., Rosi, N. L., and Jin, R. Gold-thiolate ring as a protecting motif in the $Au_{20}(SR)_{16}$ nanocluster and implications. *J. Am. Chem. Soc.*, 136:11922–11925, 2014. DOI: 10.1021/ja506802n. 34

[36] Gan, Z., Lin, Y., Luo, L., Han, G., Liu, W., Liu, Z., Yao, C., Weng, L., Liao, L., Chen, J., Liu, X., Luo, Y., Wang, C., Wei, S., and Wu, Z. Fluorescent gold nanoclusters with interlocked staples and a fully thiolate-bound kernel. *Angew. Chem. Int. Ed.*, 55:11567–11571, 2016. DOI: 10.1002/anie.201606661. 34

[37] Das, A., Li, T., Li, G., Nobusada, K., Zeng, C., Rosi, N. L., and Jin, R. Crystal structure and electronic properties of a thiolate-protected Au_{24} nanocluster. *Nanoscale*, 6:6458–6462, 2014. DOI: 10.1039/c4nr01350f. 34

[38] Liu, C., Li, T., Li, G., Nobusada, K., Zeng, C., Pang, G., Rosi, N. L., and Jin, R. Observation of body-centered cubic gold nanocluster. *Angew. Chem. Int. Ed.*, 54:9826–9829, 2015. DOI: 10.1002/anie.201502667. 34

[39] Higaki, T., Liu, C., Zhou, M., Luo, T.-Y., Rosi, N. L., and Jin, R. tailoring the structure of 58-electron gold nanoclusters: $Au_{103}S_2(S\text{-}Nap)_{41}$ and its implications. *J. Am. Chem. Soc.*, 139:9994–10001, 2017. DOI: 10.1021/jacs.7b04678. 34

[40] Zeng, C., Chen, Y., Kirschbaum, K., Appavoo, K., Sfeir, M. Y., and Jin, R. Structural patterns at all scales in a nonmetallic chiral Au_{133} $(SR)_{52}$ nanoparticle. *Sci. Adv.*, 1:e1500045, 2015. DOI: 10.1126/sciadv.1500045. 34

[41] Zeng, C., Chen, Y., Kirschbaum, K., Lambright, K. J., and Jin, R. Emergence of hierarchical structural complexities in nanoparticles and their assembly. *Science*, 354:1580–1584, 2016. DOI: 10.1126/science.aak9750. 34

[42] Zeng, C., Qian, H., Li, T., Li, G., Rosi, N. L., Yoon, B., Barnett, R. N., Whetten, R. L., Landman, U., and Jin, R. Total structure and electronic properties of the gold nanocrystal $Au_{36}(SR)_{24}$. *Angew. Chem. Int. Ed.*, 51:13114–13118, 2012. DOI: 10.1002/anie.201207098. 34

[43] Zeng, C., Li, T., Das, A., Rosi, N. L., and Jin, R. Chiral structure of thiolate-protected 28-gold-atom nanocluster determined by X-ray crystallography. *J. Am. Chem. Soc.*, 135:10011–10013, 2013. DOI: 10.1021/ja404058q. 34, 57

[44] Zeng, C., Chen, Y., Iida, K., Nobusada, K., Kirschbaum, K., Lambright, K. J., and Jin, R. Gold quantum boxes: On the periodicities and the quantum confinement in the Au_{28}, Au_{36}, Au_{44}, and Au_{52} magic series. *J. Am. Chem. Soc.*, 138:3950–3953, 2016. DOI: 10.1021/jacs.5b12747. 34, 35

[45] Zeng, C., Chen, Y., Liu, C., Nobusada, K., Rosi, N. L., and Jin, R. Gold tetrahedra coil up: Kekulé-like and double helical superstructures. *Sci. Adv.*, 1:e1500425, 2015. DOI: 10.1126/sciadv.1500425. 34

[46] Zeng, C., Liu, C., Chen, Y., Rosi, N. L., and Jin, R. Atomic structure of self-assembled monolayer of thiolates on a tetragonal Au_{92} nanocrystal. *J. Am. Chem. Soc.*, 138:8710–8713, 2016. DOI: 10.1021/jacs.6b04835. 34

[47] Sakthivel, N. A., Theivendran, S., Ganeshraj, V., Oliver, A. G., and Dass, A. Crystal structure of faradaurate-279: $Au_{279}(SPh\text{-}tBu)_{84}$ plasmonic nanocrystal molecules. *J. Am. Chem. Soc.*, 139:15450–15459, 2017. DOI: 10.1021/jacs.7b08651. 34

[48] Higaki, T., Zhou, M., Lambright, K., Kirschbaum, K., Sfeir, M. Y., and Jin, R. Sharp transition from nonmetallic Au246 to metallic Au279 with nascent surface plasmon resonance. *J. Am. Chem. Soc.*, 140:5691–5695, 2018. DOI: 10.1021/jacs.8b02487.

[49] Liao, L., Wang, C., Zhuang, S., Yan, N., Zhao, Y., Yang, Y., Li, J., Deng, H., and Wu, Z. An unprecedented kernel growth mode and layer-number-odevity-dependent properties in gold nanoclusters. *Angew. Chem. Int. Ed.*, 59:731–734, 2020. DOI: 10.1002/anie.201912090. 34

[50] McPartlin, M., Mason, R., and Malatesta, L. Novel cluster complexes of gold(0)-Gold(I). *J. Chem. Soc. D*, pages 334–334, 1969. DOI: 10.1039/c29690000334. 35

[51] Mingos, D. M. P. Molecular-orbital calculations on cluster compounds of gold. *J. Chem. Soc., Dalton Trans.*, pages 1163–1169, 1976. DOI: 10.1039/dt9760001163. 35

[52] Briant, C. E., Theobald, B. R. C., White, J. W., Bell, L. K., Mingos, D. M. P., and Welch, A. J. Synthesis and X-ray structural characterization of the centred icosahedral gold cluster compound $[Au_{13}(PMe_2Ph)_{10}Cl_2](PF_6)_3$, the realization of a theoretical prediction. *J. Chem. Soc., Chem. Commun.*, pages 201–202, 1981. DOI: 10.1039/C39810000201.

[53] Teo, B. K., Shi, X., and Zhang, H. Pure gold cluster of 1:9:9:1:9:9:1 layered structure: A novel 39-metal-atom cluster $[(Ph_3P)_{14}Au_{39}Cl_6]Cl_2$ with an interstitial gold atom in a hexagonal antiprismatic cage. *J. Am. Chem. Soc.*, 114:2743–2745, 1992. DOI: 10.1021/ja00033a073. 35

[54] Chen, J., Zhang, Q.-F., Bonaccorso, T. A., Williard, P. G., and Wang, L.-S. Controlling gold nanoclusters by diphospine ligands. *J. Am. Chem. Soc.*, 136:92–95, 2014. DOI: 10.1021/ja411061e. 35

[55] Shichibu, Y., Negishi, Y., Watanabe, T., Chaki, N. K., Kawaguchi, H., and Tsukuda, T. Biicosahedral gold clusters $[Au_{25}(PPh_3)_{10}(SC_nH_{2n+1})_5Cl_2]^{2+}$ ($n = 2 - 18$): A stepping stone to cluster-assembled materials. *J. Phys. Chem. C*, 111:7845–7847, 2007. DOI: 10.1021/jp073101t. 35

[56] Song, Y., Wang, S., Zhang, J., Kang, X., Chen, S., Li, P., Sheng, H., and Zhu, M. Crystal structure of selenolate-protected $Au_{24}(SeR)_{20}$ nanocluster. *J. Am. Chem. Soc.*, 136:2963–2965, 2014. DOI: 10.1021/ja4131142. 35

[57] Lei, Z., Wan, X.-K., Yuan, S.-F., Guan, Z.-J., and Wang, Q.-M. Alkynyl approach toward the protection of metal nanoclusters. *Acc. Chem. Res.*, 51:2465–2474, 2018. DOI: 10.1021/acs.accounts.8b00359. 35

[58] Bhattarai, B., Zaker, Y., Atnagulov, A., Yoon, B., Landman, U., and Bigioni, T. P. Chemistry and structure of silver molecular nanoparticles. *Acc. Chem. Res.*, 51:3104–3113, 2018. DOI: 10.1021/acs.accounts.8b00445. 35, 37

[59] Yan, J., Teo, B. K., and Zheng, N. Surface chemistry of atomically precise coinage-metal nanoclusters: From structural control to surface reactivity and catalysis. *Acc. Chem. Res.*, 51:3084–3093, 2018. DOI: 10.1021/acs.accounts.8b00371. 35

[60] Chakraborty, I., Kurashige, W., Kanehira, K., Gell, L., Häkkinen, H., Negishi, Y., and Pradeep, T. $Ag_{44}(SeR)_{30}$: A hollow cage silver cluster with selenolate protection. *J. Phys. Chem. Lett.*, 4:3351–3355, 2013. DOI: 10.1021/jz401879c. 35

[61] Bootharaju, M. S., Joshi, C. P., Alhilaly, M. J., and Bakr, O. M. Switching a nanocluster core from hollow to nonhollow. *Chem. Mater.*, 28:3292–3297, 2016. DOI: 10.1021/acs.chemmater.5b05008. 35

[62] Chai, J., Yang, S., Lv, Y., Chen, T., Wang, S., Yu, H., and Zhu, M. A unique pair: Ag_{40} and Ag_{46} nanoclusters with the same surface but different cores for structure—property correlation. *J. Am. Chem. Soc.*, 140:15582–15585, 2018. DOI: 10.1021/jacs.8b09162. 35

[63] Yang, H., Lei, J., Wu, B., Wang, Y., Zhou, M., Xia, A., Zheng, L., and Zheng, N. Crystal structure of a luminescent thiolated Ag nanocluster with an octahedral Ag_6^{4+} core. *Chem. Commun.*, 49:300–302, 2013. DOI: 10.1039/c2cc37347e. 35, 36

[64] Desireddy, A., Conn, B. E., Guo, J., Yoon, B., Barnett, R. N., Monahan, B. M., Kirschbaum, K., Griffith, W. P., Whetten, R. L., Landman, U., and Bigioni, T. P. Ultrastable silver nanoparticles. *Nature*, 501:399–402, 2013. DOI: 10.1038/nature12523. 36, 45

[65] Yang, H., Wang, Y., Huang, H., Gell, L., Lehtovaara, L., Malola, S., Häkkinen, H., and Zheng, N. All-thiol-stabilized Ag_{44} and $Au_{12}Ag_{32}$ nanoparticles with single-crystal structures. *Nat. Commun.*, 4:2422, 2013. DOI: 10.1038/ncomms3422. 36

[66] AbdulHalim, L. G., Bootharaju, M. S., Tang, Q., Del Gobbo, S., AbdulHalim, R. G., Eddaoudi, M., Jiang, D.-E., and Bakr, O. M. $Ag_{29}(BDT)_{12}(TPP)_4$: A tetravalent nanocluster. *J. Am. Chem. Soc.*, 137:11970–11975, 2015. DOI: 10.1021/jacs.5b04547. 36

[67] Joshi, C. P., Bootharaju, M. S., Alhilaly, M. J., and Bakr, O. M. $[Ag_{25}(SR)_{18}]^-$: The "golden" silver nanoparticle. *J. Am. Chem. Soc.*, 137:11578–11581, 2015. DOI: 10.1021/jacs.5b07088. 36, 37

[68] Kang, X. and Zhu, M. Transformation of atomically precise nanoclusters by ligand-exchange. *Chem. Mater.*, 31:9939–9969, 2019. DOI: 10.1021/acs.chemmater.9b03674. 37

[69] Nguyen, T. A. D., Jones, Z. R., Goldsmith, B. R., Buratto, W. R., Wu, G., Scott, S. L., and Hayton, T. W. A Cu_{25} nanocluster with partial Cu(0) character. *J. Am. Chem. Soc.*, 137:13319–13324, 2015. DOI: 10.1021/jacs.5b07574. 37

[70] Tran, N. T., Powell, D. R., and Dahl, L. F. Nanosized $Pd_{145}(CO)_x(PEt_3)_{30}$ containing a capped three-shell 145-atom metal-core geometry of pseudo icosahedral symmetry. *Angew. Chem. Int. Ed.*, 39:4121–4125, 2000. DOI: 10.1002/1521-3773(20001117)39:22%3C4121::AID-ANIE4121%3E3.0.CO;2-A. 37

[71] Erickson, J. D., Mednikov, E. G., Ivanov, S. A., and Dahl, L. F. Isolation and structural characterization of a mackay 55-metal-atom two-shell icosahedron of pseudo-Ih symmetry, $Pd_{55}L_{12}(\mu3\text{-}CO)_{20}$ (L = PR3, R = isopropyl): Comparative analysis with interior two-shell icosahedral geometries in capped three-shell Pd_{145}, Pt-centered four-shell Pd-Pt M165, and four-Shell Au_{133} nanoclusters. *J. Am. Chem. Soc.*, 138:1502–1505, 2016. DOI: 10.1021/jacs.5b13076. 37

[72] Liao, L., Zhou, S., Dai, Y., Liu, L., Yao, C., Fu, C., Yang, J., and Wu, Z. Mono-mercury doping of Au_{25} and the HOMO/LUMO energies evaluation employing differential pulse voltammetry. *J. Am. Chem. Soc.*, 137:9511–9514, 2015. DOI: 10.1021/jacs.5b03483. 37, 48, 58, 59, 60

[73] Yao, C., Lin, Y.-j., Yuan, J., Liao, L., Zhu, M., Weng, L.-h., Yang, J., and Wu, Z. Mono-cadmium vs. mono-mercury doping of Au_{25} nanoclusters. *J. Am. Chem. Soc.*, 137:15350–15353, 2015. DOI: 10.1021/jacs.5b09627. 37

[74] Kauffman, D. R., Alfonso, D., Matranga, C., Qian, H., and Jin, R. A quantum alloy: The ligand-protected $Au_{25--x}Ag_x(SR)_{18}$ cluster. *J. Phys. Chem. C*, 117:7914–7923, 2013. DOI: 10.1021/jp4013224. 37

[75] Yan, N., Liao, L. W., Yuan, J. Y., Lin, Y. J., Weng, L. H., Yang, J. L., and Wu, Z. K. Bimetal doping in nanoclusters: Synergistic or counteractive?. *Chem. Mater.*, 28:8240–8247, 2016. DOI: 10.1021/acs.chemmater.6b03132. 37, 48

[76] Yan, J., Su, H., Yang, H., Hu, C., Malola, S., Lin, S., Teo, B. K., Häkkinen, H., and Zheng, N. Asymmetric synthesis of chiral bimetallic $[Ag_{28}Cu_{12}(SR)_{24}]^{4-}$ nanoclusters via ion pairing. *J. Am. Chem. Soc.*, 138:12751–12754, 2016. DOI: 10.1021/jacs.6b08100. 37

[77] Jones, C. G., Martynowycz, M. W., Hattne, J., Fulton, T. J., Stoltz, B. M., Rodriguez, J. A., Nelson, H. M., and Gonen, T. The cryoEM method microED as a powerful tool for small molecule structure determination. *ACS Cent. Sci.*, 4:1587–1592, 2018. DOI: 10.1021/acscentsci.8b00760. 37

[78] Vergara, S., Lukes, D. A., Martynowycz, M. W., Santiago, U., Plascencia-Villa, G., Weiss, S. C., de la Cruz, M. J., Black, D. M., Alvarez, M. M., López-Lozano, X., Barnes, C. O., Lin, G., Weissker, H.-C., Whetten, R. L., Gonen, T., Yacaman, M. J., and Calero, G. MicroED structure of $Au_{146}(p-MBA)_{57}$ at subatomic resolution reveals a twinned FCC cluster. *J. Phys. Chem. Lett.*, 8:5523–5530, 2017. DOI: 10.1021/acs.jpclett.7b02621. 37, 38

[79] Zhang, P. X-ray spectroscopy of gold-thiolate nanoclusters. *J. Phys. Chem. C*, 118:25291–25299, 2014. DOI: 10.1021/jp507739u. 38

[80] Yamazoe, S. and Tsukuda, T. X-ray absorption spectroscopy on atomically precise metal clusters. *Bull. Chem. Soc. Jpn.*, 92:193–204, 2019. DOI: 10.1246/bcsj.20180282. 38

[81] Li, Y., Cheng, H., Yao, T., Sun, Z., Yan, W., Jiang, Y., Xie, Y., Sun, Y., Huang, Y., Liu, S., Zhang, J., Xie, Y., Hu, T., Yang, L., Wu, Z., and Wei, S. Hexane-driven icosahedral to cuboctahedral structure transformation of gold nanoclusters. *J. Am. Chem. Soc.*, 134:17997–18003, 2012. DOI: 10.1021/ja306923a. 38, 39

[82] MacDonald, M. A., Zhang, P., Qian, H., and Jin, R. Site-specific and size-dependent bonding of compositionally precise gold-thiolate nanoparticles from X-ray spectroscopy. *J. Phys. Chem. Lett.*, 1:1821–1825, 2010. DOI: 10.1021/jz100547q. 38

[83] Lopez-Acevedo, O., Akola, J., Whetten, R. L., Grönbeck, H., and Häkkinen, H. Structure and bonding in the ubiquitous icosahedral metallic gold cluster $Au_{144}(SR)_{60}$. *J. Phys. Chem. C*, 113:5035–5038, 2009. DOI: 10.1021/jp8115098. 38

[84] Omoda, T., Takano, S., Yamazoe, S., Koyasu, K., Negishi, Y., and Tsukuda, T. An $Au_{25}(SR)_{18}$ cluster with a face-centered cubic core. *J. Phys. Chem. C*, 122:13199–13204, 2018. DOI: 10.1021/acs.jpcc.8b03841. 38, 39

[85] MacDonald, M. A., Chevrier, D. M., Zhang, P., Qian, H., and Jin, R. The structure and bonding of $Au_{25}(SR)_{18}$ nanoclusters from EXAFS: The interplay of metallic and molecular behavior. *J. Phys. Chem. C*, 115:15282–15287, 2011. DOI: 10.1021/jp204922m. 39

[86] Yamazoe, S., Takano, S., Kurashige, W., Yokoyama, T., Nitta, K., Negishi, Y., and Tsukuda, T. Hierarchy of bond stiffnesses within icosahedral-based gold clusters protected by thiolates. *Nat. Commun.*, 7:10414, 2016. DOI: 10.1038/ncomms10414. 39, 40

[87] Christensen, S. L., MacDonald, M. A., Chatt, A., Zhang, P., Qian, H., and Jin, R. Dopant location, local structure, and electronic properties of $Au_{24}Pt(SR)_{18}$ nanoclusters. *J. Phys. Chem. C*, 116:26932–26937, 2012. DOI: 10.1021/jp310183x. 40

[88] Negishi, Y., Kurashige, W., Kobayashi, Y., Yamazoe, S., Kojima, N., Seto, M., and Tsukuda, T. Formation of a Pd@Au_{12} superatomic core in $Au_{24}Pd_1(SC_{12}H_{25})_{18}$ probed by ^{197}Au Mössbauer and Pd K-edge EXAFS spectroscopy. *J. Phys. Chem. Lett.*, 4:3579–3583, 2013. DOI: 10.1021/jz402030n. 40

[89] Tian, S., Liao, L., Yuan, J., Yao, C., Chen, J., Yang, J., and Wu, Z. Structures and magnetism of mono-palladium and mono-platinum doped $Au_{25}(PET)_{18}$ nanoclusters. *Chem. Commun.*, 52:9873–9876, 2016. DOI: 10.1039/c6cc02698b. 40, 60

[90] Tofanelli, M. A., Ni, T. W., Phillips, B. D., and Ackerson, C. J. Crystal structure of the $PdAu_{24}(SR)_{18}^0$ superatom. *Inorg. Chem.*, 55:999–1001, 2016. DOI: 10.1021/acs.inorgchem.5b02106. 40

[91] Agrachev, M., Ruzzi, M., Venzo, A., and Maran, F. Nuclear and electron magnetic resonance spectroscopies of atomically precise gold nanoclusters. *Acc. Chem. Res.*, 52:44–52, 2019. DOI: 10.1021/acs.accounts.8b00495. 40

[92] Salassa, G. and Burgi, T. NMR spectroscopy: A potent tool for studying monolayer-protected metal nanoclusters. *Nanoscale Horiz.*, 3:457–463, 2018. DOI: 10.1039/c8nh00058a. 40

[93] Wu, Z., Gayathri, C., Gil, R. R., and Jin, R. Probing the structure and charge state of glutathione-capped $Au_{25}(SG)_{18}$ clusters by NMR and mass spectrometry. *J. Am. Chem. Soc.*, 131:6535–6542, 2009. DOI: 10.1021/ja900386s. 40

[94] Chen, Y., Zeng, C., Liu, C., Kirschbaum, K., Gayathri, C., Gil, R. R., Rosi, N. L., and Jin, R. Crystal structure of barrel-shaped chiral $Au_{130}(p\text{-MBT})_{50}$ nanocluster. *J. Am. Chem. Soc.*, 137:10076–10079, 2015. DOI: 10.1021/jacs.5b05378. 40, 41

[95] Xia, N., Yang, J., and Wu, Z. Fast, high-yield synthesis of amphiphilic Ag nanoclusters and the sensing of Hg^{2+} in environmental samples. *Nanoscale*, 7:10013–10020, 2015. DOI: 10.1039/c5nr00705d. 40

[96] Wu, Z. and Jin, R. Stability of the two Au-S binding modes in $Au_{25}(SG)_{18}$ nanoclusters probed by NMR and optical spectroscopy. *ACS Nano*, 3:2036–2042, 2009. DOI: 10.1021/nn9004999. 40, 41

[97] Parker, J. F., Fields-Zinna, C. A., and Murray, R. W. The story of a monodisperse gold nanoparticle: $Au_{25}L_{18}$. *Acc. Chem. Res.*, 43:1289–1296, 2010. DOI: 10.1021/ar100048c. 41

[98] Venzo, A., Antonello, S., Gascón, J. A., Guryanov, I., Leapman, R. D., Perera, N. V., Sousa, A., Zamuner, M., Zanella, A., and Maran, F. Effect of the charge state ($z = -1, 0, +1$) on the nuclear magnetic resonance of monodisperse $Au_{25}[S(CH_2)_2Ph]_{18}^z$ clusters. *Anal. Chem.*, 83:6355–6362, 2011. DOI: 10.1021/ac2012653. 41

[99] Zeng, C., Weitz, A., Withers, G., Higaki, T., Zhao, S., Chen, Y., Gil, R. R., Hendrich, M., and Jin, R. Controlling magnetism of $Au_{133}(TBBT)_{52}$ nanoclusters at single electron level and implication for nonmetal to metal transition. *Chem. Sci.*, 10:9684–9691, 2019. DOI: 10.1039/c9sc02736j. 41

[100] Qian, H., Zhu, M., Gayathri, C., Gil, R. R., and Jin, R. Chirality in gold nanoclusters probed by NMR spectroscopy. *ACS Nano*, 5:8935–8942, 2011. DOI: 10.1021/nn203113j. 41

[101] Schaaff, T. G., Shafigullin, M. N., Khoury, J. T., Vezmar, I., Whetten, R. L., Cullen, W. G., First, P. N., Gutiérrez-Wing, C., Ascensio, J., and Jose-Yacamán, M. J. Isolation of smaller nanocrystal Au molecules: Robust quantum effects in optical spectra. *J. Phys. Chem. B*, 101:7885–7891, 1997. DOI: 10.1021/jp971438x. 42, 43

[102] Tracy, J. B., Kalyuzhny, G., Crowe, M. C., Balasubramanian, R., Choi, J.-P., and Murray, R. W. Poly(ethylene glycol) ligands for high-resolution nanoparticle mass spectrometry. *J. Am. Chem. Soc.*, 129:6706–6707, 2007. DOI: 10.1021/ja071042r. 42, 50

[103] Tsunoyama, H., Negishi, Y., and Tsukuda, T. Chromatographic isolation of missing, Au_{55} clusters protected by alkanethiolates. *J. Am. Chem. Soc.*, 128:6036–6037, 2006. DOI: 10.1021/ja061659t. 42

[104] Dass, A., Stevenson, A., Dubay, G. R., Tracy, J. B., and Murray, R. W. Nanoparticle MALDI-TOF mass spectrometry without fragmentation: $Au_{25}(SCH_2CH_2Ph)_{18}$ and mixed monolayer $Au_{25}(SCH_2CH_2Ph)_{18-x}(L)_x$. *J. Am. Chem. Soc.*, 130:5940–5946, 2008. DOI: 10.1021/ja710323t. 42, 44

[105] Negishi, Y., Kurashige, W., Niihori, Y., Iwasa, T., and Nobusada, K. Isolation, structure, and stability of a dodecanethiolate-protected Pd_1Au_{24} cluster. *Phys. Chem. Chem. Phys.*, 12:6219–6225, 2010. DOI: 10.1039/b927175a. 42

[106] Lu, Y., Jiang, Y., Gao, X., and Chen, W. Charge state-dependent catalytic activity of $[Au_{25}(SC_{12}H_{25})_{18}]$ nanoclusters for the two-electron reduction of dioxygen to hydrogen peroxide. *Chem. Commun.*, 50:8464–8467, 2014. DOI: 10.1039/c4cc01841a. 42

[107] Nimmala, P. R. and Dass, A. $Au_{36}(SPh)_{23}$ nanomolecules. *J. Am. Chem. Soc.*, 133:9175–9177, 2011. DOI: 10.1021/ja201685f. 42

[108] Vergara, S., Santiago, U., Kumara, C., Alducin, D., Whetten, R. L., Jose Yacaman, M., Dass, A., and Ponce, A. Synthesis, mass spectrometry, and atomic structural analysis of $Au_{\sim2000}(SR)_{\sim290}$ nanoparticles. *J. Phys. Chem. C*, 122:26733–26738, 2018. DOI: 10.1021/acs.jpcc.8b08531. 42, 48

[109] Schaaff, T. G., Knight, G., Shafigullin, M. N., Borkman, R. F., and Whetten, R. L. Isolation and selected properties of a 10.4 kDa gold: Glutathione cluster compound. *J. Phys. Chem. B*, 102:10643–10646, 1998. DOI: 10.1021/jp9830528. 43

[110] Schaaff, T. G. and Whetten, R. L. Giant gold-glutathione cluster compounds: Intense optical activity in metal-based transitions. *J. Phys. Chem. B*, 104:2630–2641, 2000. DOI: 10.1021/jp993691y. 43

[111] Negishi, Y., Takasugi, Y., Sato, S., Yao, H., Kimura, K., and Tsukuda, T. Magic-numbered Au_n clusters protected by glutathione monolayers ($n = 18, 21, 25, 28, 32, 39$): Isolation and spectroscopic characterization. *J. Am. Chem. Soc.*, 126:6518–6519, 2004. DOI: 10.1021/ja0483589. 43

[112] Negishi, Y., Nobusada, K., and Tsukuda, T. Glutathione-protected gold clusters revisited: Bridging the gap between gold(I)-thiolate complexes and thiolate-protected gold nanocrystals. *J. Am. Chem. Soc.*, 127:5261–5270, 2005. DOI: 10.1021/ja042218h. 43, 44, 48

[113] Qian, H., Zhu, Y., and Jin, R. Atomically precise gold nanocrystal molecules with surface plasmon resonance. *Proc. Natl. Acad. Sci.*, 109:696–700, 2012. DOI: 10.1073/pnas.1115307109. 44, 45

[114] Wu, Z., Lanni, E., Chen, W., Bier, M. E., Ly, D., and Jin, R. High yield, large scale synthesis of thiolate-protected Ag_7 clusters. *J. Am. Chem. Soc.*, 131:16672–16673, 2009. DOI: 10.1021/ja907627f. 44, 45, 46

[115] Wei, W., Lu, Y., Chen, W., and Chen, S. One-pot synthesis, photoluminescence, and electrocatalytic properties of subnanometer-sized copper clusters. *J. Am. Chem. Soc.*, 133:2060–2063, 2011. DOI: 10.1021/ja109303z. 45, 51

[116] Tanaka, S.-I., Miyazaki, J., Tiwari, D. K., Jin, T., and Inouye, Y. Fluorescent platinum nanoclusters: Synthesis, purification, characterization, and application to bioimaging. *Angew. Chem. Int. Ed.*, 50:431–435, 2011. DOI: 10.1002/anie.201004907. 45

[117] Chen, J., Liu, L., Weng, L., Lin, Y., Liao, L., Wang, C., Yang, J., and Wu, Z. Synthesis and properties evolution of a family of tiara-like phenylethanethiolated palladium nanoclusters. *Sci. Rep.*, 5:16628, 2015. DOI: 10.1038/srep16628. 45

[118] Negishi, Y., Iwai, T., and Ide, M. Continuous modulation of electronic structure of stable thiolate-protected Au_{25} cluster by Ag doping. *Chem. Commun.*, 46:4713–4715, 2010. DOI: 10.1039/c0cc01021a. 45

[119] Qian, H., Jiang, D.-E., Li, G., Gayathri, C., Das, A., Gil, R. R., and Jin, R. Monoplatinum doping of gold nanoclusters and catalytic application. *J. Am. Chem. Soc.*, 134:16159–16162, 2012. DOI: 10.1021/ja307657a. 45

[120] Yao, C., Chen, J., Li, M.-B., Liu, L., Yang, J., and Wu, Z. Adding two active silver atoms on Au_{25} nanoparticle. *Nano Lett.*, 15:1281–1287, 2015. DOI: 10.1021/nl504477t. 45

[121] Negishi, Y., Chaki, N. K., Shichibu, Y., Whetten, R. L., and Tsukuda, T. Origin of magic stability of thiolated gold clusters: A case study on $Au_{25}(SC_6H_{13})_{18}$. *J. Am. Chem. Soc.*, 129:11322–11323, 2007. DOI: 10.1021/ja073580+. 45

[122] Tracy, J. B., Crowe, M. C., Parker, J. F., Hampe, O., Fields-Zinna, C. A., Dass, A., and Murray, R. W. Electrospray ionization mass spectrometry of uniform and mixed monolayer nanoparticles: $Au_{25}[S(CH_2)_2Ph]_{18}$ and $Au_{25}[S(CH_2)_2Ph]_{18-x}(SR)_x$. *J. Am. Chem. Soc.*, 129:16209–16215, 2007. DOI: 10.1021/ja076621a. 45

[123] Dharmaratne, A. C., Krick, T., and Dass, A. Nanocluster size evolution studied by mass spectrometry in room temperature $Au_{25}(SR)_{18}$ synthesis. *J. Am. Chem. Soc.*, 131:13604–13605, 2009. DOI: 10.1021/ja906087a. 45

[124] Wu, Z., Suhan, J., and Jin, R. One-pot synthesis of atomically monodisperse, thiol-functionalized Au_{25} nanoclusters. *J. Mater. Chem.*, 19:622–626, 2009. DOI: 10.1039/b815983a. 45, 52, 53

[125] Zeng, C., Liu, C., Pei, Y., and Jin, R. Thiol ligand-induced transformation of $Au_{38}(SC_2H_4Ph)_{24}$ to $Au_{36}(SPh\ t\ Bu)_{24}$. *ACS Nano*, 7:6138–6145, 2013. DOI: 10.1021/nn401971g. 45

[126] Yao, Q., Yuan, X., Fung, V., Yu, Y., Leong, D. T., Jiang, D.-E., and Xie, J. Understanding seed-mediated growth of gold nanoclusters at molecular level. *Nat. Commun.*, 8:927, 2017. DOI: 10.1038/s41467-017-00970-1. 45

[127] Azubel, M., Koivisto, J., Malola, S., Bushnell, D., Hura, G. L., Koh, A. L., Tsunoyama, H., Tsukuda, T., Pettersson, M., Häkkinen, H., and Kornberg, R. D. Electron microscopy of gold nanoparticles at atomic resolution. *Science*, 345:909–912, 2014. DOI: 10.1126/science.1251959. 45, 47

[128] Azubel, M., Koh, A. L., Koyasu, K., Tsukuda, T., and Kornberg, R. D. Structure determination of a water-soluble 144-gold atom particle at atomic resolution by aberration-corrected electron microscopy. *ACS Nano*, 11:11866–11871, 2017. DOI: 10.1021/acsnano.7b06051. 47

[129] Liao, L., Chen, J., Wang, C., Zhuang, S., Yan, N., Yao, C., Xia, N., Li, L., Bao, X., and Wu, Z. Transition-sized Au_{92} nanoparticle bridging non-fcc-structured gold nanoclusters and fcc-structured gold nanocrystals. *Chem. Commun.*, 52:12036–12039, 2016. DOI: 10.1039/c6cc06108g. 47

[130] Kumara, C., Zuo, X., Ilavsky, J., Chapman, K. W., Cullen, D. A., and Dass, A. Super-stable, highly monodisperse plasmonic faradaurate-500 nanocrystals with 500 gold atoms: $Au_{\sim500}(SR)_{\sim120}$. *J. Am. Chem. Soc.*, 136:7410–7417, 2014. DOI: 10.1021/ja502327a. 48

[131] Kumara, C., Zuo, X., Cullen, D. A., and Dass, A. Faradaurate-940: Synthesis, mass spectrometry, electron microscopy, high-energy X-ray diffraction, and X-ray scattering study of $Au_{\sim940\pm20}(SR)_{\sim160\pm4}$ nanocrystals. *ACS Nano*, 8:6431–6439, 2014. DOI: 10.1021/nn501970v. 48

[132] Kumara, C., Hoque, M. M., Zuo, X., Cullen, D. A., Whetten, R. L., and Dass, A. Isolation of a 300 kDa, Au~ 1400 gold compound, the standard 3.6 nm capstone to a series of plasmonic nanocrystals protected by aliphatic-like thiolates. *J. Phys. Chem. Lett.*, 9:6825–6832, 2018. DOI: 10.1021/acs.jpclett.8b02993. 48

[133] Yang, H., Wang, Y., Chen, X., Zhao, X., Gu, L., Huang, H., Yan, J., Xu, C., Li, G., Wu, J., Edwards, A. J., Dittrich, B., Tang, Z., Wang, D., Lehtovaara, L., Häkkinen, H., and Zheng, N. Plasmonic twinned silver nanoparticles with molecular precision. *Nat. Commun.*, 7:12809, 2016. DOI: 10.1038/ncomms12809. 48

[134] Shichibu, Y., Negishi, Y., Tsunoyama, H., Kanehara, M., Teranishi, T., and Tsukuda, T. Extremely high stability of glutathionate-protected Au_{25} clusters against core etching. *Small*, 3:835–839, 2007. DOI: 10.1002/smll.200600611. 48

[135] Liao, L., Zhuang, S., Yao, C., Yan, N., Chen, J., Wang, C., Xia, N., Liu, X., Li, M.-B., Lo, L., Bao, X., and Wu, Z. Structure of chiral $Au_{44}(2,4\text{-DMBT})_{26}$ nanocluster with an 18-electron shell closure. *J. Am. Chem. Soc.*, 138:10425–10428, 2016. DOI: 10.1021/jacs.6b07178. 48, 49

[136] Vilar-Vidal, N., Blanco, M. C., López-Quintela, M. A., Rivas, J., and Serra, C. Electrochemical synthesis of very stable photoluminescent copper clusters. *J. Phys. Chem. C*, 114:15924–15930, 2010. DOI: 10.1021/jp911380s. 48

[137] Hostetler, M. J., Wingate, J. E., Zhong, C.-J., Harris, J. E., Vachet, R. W., Clark, M. R., Londono, J. D., Green, S. J., Stokes, J. J., Wignall, G. D., Glish, G. L., Porter, M. D., Evans, N. D., and Murray, R. W. Alkanethiolate gold cluster molecules with core diameters from 1.5–5.2 nm: Core and monolayer properties as a function of core size. *Langmuir*, 14:17–30, 1998. DOI: 10.1021/la970588w. 48

[138] Levi-Kalisman, Y., Jadzinsky, P. D., Kalisman, N., Tsunoyama, H., Tsukuda, T., Bushnell, D. A., and Kornberg, R. D. Synthesis and characterization of $Au_{102}(p\text{-MBA})_{44}$ nanoparticles. *J. Am. Chem. Soc.*, 133:2976–2982, 2011. DOI: 10.1021/ja109131w. 48

[139] Zhu, Y., Qian, H., and Jin, R. An atomic-level strategy for unraveling gold nanocatalysis from the perspective of $Au_n(SR)_m$ nanoclusters. *Chem. Eur. J.*, 16:11455–11462, 2010. DOI: 10.1002/chem.201001086. 49

[140] Li, G., Jiang, D.-E., Liu, C., Yu, C., and Jin, R. Oxide-supported atomically precise gold nanocluster for catalyzing sonogashira cross-coupling. *J. Catal.*, 306:177–183, 2013. DOI: 10.1016/j.jcat.2013.06.017. 49

[141] Kurashige, W., Yamaguchi, M., Nobusada, K., and Negishi, Y. Ligand-induced stability of gold nanoclusters: Thiolate vs. selenolate. *J. Phys. Chem. Lett.*, 3:2649–2652, 2012. DOI: 10.1021/jz301191t. 49, 50

[142] Creighton, J. A. and Eadon, D. G. Ultraviolet—visible absorption spectra of the colloidal metallic elements. *J. Chem. Soc., Faraday Trans.*, 87:3881–3891, 1991. DOI: 10.1039/ft9918703881. 50

[143] Schwartzberg, A. M. and Zhang, J. Z. Novel optical properties and emerging applications of metal nanostructures. *J. Phys. Chem. C*, 112:10323–10337, 2008. DOI: 10.1021/jp801770w. 50

[144] Zhu, M., Lanni, E., Garg, N., Bier, M. E., and Jin, R. Kinetically controlled, high-yield synthesis of Au_{25} clusters. *J. Am. Chem. Soc.*, 130:1138–1139, 2008. DOI: 10.1021/ja0782448. 50

[145] Bakr, O. M., Amendola, V., Aikens, C. M., Wenseleers, W., Li, R., Dal Negro, L., Schatz, G. C., and Stellacci, F. Silver nanoparticles with broad multiband linear optical absorption. *Angew. Chem. Int. Ed.*, 48:5921–5926, 2009. DOI: 10.1002/anie.200900298. 51

[146] Devadas, M. S., Bairu, S., Qian, H., Sinn, E., Jin, R., and Ramakrishna, G. Temperature-dependent optical absorption properties of monolayer-protected Au_{25} and Au_{38} clusters. *J. Phys. Chem. Lett.*, 2:2752–2758, 2011. DOI: 10.1021/jz2012897. 51, 52

[147] Weissker, H. C., Escobar, H. B., Thanthirige, V. D., Kwak, K., Lee, D., Ramakrishna, G., Whetten, R. L., and López-Lozano, X. Information on quantum states pervades the visible spectrum of the ubiquitous $Au_{144}(SR)_{60}$ gold nanocluster. *Nat. Commun.*, 5:3785, 2014. DOI: 10.1038/ncomms4785. 51

[148] Sementa, L., Barcaro, G., Dass, A., Stener, M., and Fortunelli, A. Designing ligand-enhanced optical absorption of thiolated gold nanoclusters. *Chem. Commun.*, 51:7935–7938, 2015. DOI: 10.1039/c5cc01951f. 52

[149] Negishi, Y., Nakazaki, T., Malola, S., Takano, S., Niihori, Y., Kurashige, W., Yamazoe, S., Tsukuda, T., and Häkkinen, H. A critical size for emergence of nonbulk electronic

and geometric structures in dodecanethiolate-protected Au clusters. *J. Am. Chem. Soc.*, 137:1206–1212, 2015. DOI: 10.1021/ja5109968. 52

[150] Yu, H., Rao, B., Jiang, W., Yang, S., and Zhu, M. The photoluminescent metal nanoclusters with atomic precision. *Coord. Chem. Rev.*, 378:595–617, 2019. DOI: 10.1016/j.ccr.2017.12.005. 53

[151] Zhang, L. and Wang, E. Metal nanoclusters: New fluorescent probes for sensors and bioimaging. *Nano Today*, 9:132–157, 2014. DOI: 10.1016/j.nantod.2014.02.010. 53

[152] Yu, P., Wen, X., Toh, Y.-R., and Tang, J. Temperature-dependent fluorescence in Au_{10} nanoclusters. *J. Phys. Chem. C*, 116:6567–6571, 2012. DOI: 10.1021/jp2120077. 53

[153] Wen, X., Yu, P., Toh, Y.-R., and Tang, J. Structure-correlated dual fluorescent bands in BSA-protected Au_{25} nanoclusters. *J. Phys. Chem. C*, 116:11830–11836, 2012. DOI: 10.1021/jp303530h. 53, 54, 55

[154] Wen, X., Yu, P., Toh, Y.-R., Hsu, A.-C., Lee, Y.-C., and Tang, J. Fluorescence dynamics in BSA-protected Au_{25} nanoclusters. *J. Phys. Chem. C*, 116:19032–19038, 2012. DOI: 10.1021/jp305902w. 54

[155] Kang, X., Chong, H., and Zhu, M. $Au_{25}(SR)_{18}$: The captain of the great nanocluster ship. *Nanoscale*, 10:10758–10834, 2018. DOI: 10.1039/c8nr02973c. 54

[156] Hartland, G. V. Optical studies of dynamics in noble metal nanostructures. *Chem. Rev.*, 111:3858–3887, 2011. DOI: 10.1021/cr1002547. 54

[157] Yau, S. H., Varnavski, O., and Goodson, T., III An ultrafast look at Au nanoclusters. *Acc. Chem. Res.*, 46:1506–1516, 2013. DOI: 10.1021/ar300280w. 54

[158] Qian, H., Sfeir, M. Y., and Jin, R. Ultrafast relaxation dynamics of $[Au_{25}(SR)_{18}]^q$ nanoclusters: Effects of charge state. *J. Phys. Chem. C*, 114:19935–19940, 2010. DOI: 10.1021/jp107915x. 54, 56

[159] Green, T. D. and Knappenberger, K. L. Relaxation dynamics of $Au_{25}L_{18}$ nanoclusters studied by femtosecond time-resolved near infrared transient absorption spectroscopy. *Nanoscale*, 4:4111–4118, 2012. DOI: 10.1039/c2nr31080e. 54

[160] Zhou, M., Higaki, T., Hu, G., Sfeir, M. Y., Chen, Y., Jiang, D.-E., and Jin, R. Three-orders-of-magnitude variation of carrier lifetimes with crystal phase of gold nanoclusters. *Science*, 364:279–282, 2019. DOI: 10.1126/science.aaw8007. 56

[161] Link, S., El-Sayed, M. A., Gregory Schaaff, T., and Whetten, R. L. Transition from nanoparticle to molecular behavior: A femtosecond transient absorption study of a size-selected 28 atom gold cluster. *Chem. Phys. Lett.*, 356:240–246, 2002. DOI: 10.1016/s0009-2614(02)00306-8. 56

[162] Higaki, T., Zhou, M., Lambright, K. J., Kirschbaum, K., Sfeir, M. Y., and Jin, R. Sharp transition from nonmetallic Au_{246} to metallic Au_{279} with nascent surface plasmon resonance. *J. Am. Chem. Soc.*, 140:5691–5695, 2018. DOI: 10.1021/jacs.8b02487. 56

[163] Nieto-Ortega, B. and Bürgi, T. Vibrational properties of thiolate-protected gold nanoclusters. *Acc. Chem. Res.*, 51:2811–2819, 2018. DOI: 10.1021/acs.accounts.8b00376. 57

[164] Gautier, C. and Bürgi, T. Chiral gold nanoparticles. *ChemPhysChem*, 10:483–492, 2009. DOI: 10.1002/cphc.200800709. 57

[165] Knoppe, S. and Bürgi, T. Chirality in thiolate-protected gold clusters. *Acc. Chem. Res.*, 47:1318–1326, 2014. DOI: 10.1021/ar400295d.

[166] Noguez, C. and Garzón, I. L. Optically active metal nanoparticles. *Chem. Soc. Rev.*, 38:757–771, 2009. DOI: 10.1039/b800404h. 57

[167] Pelayo, J. J., Whetten, R. L., and Garzón, I. L. Geometric quantification of chirality in ligand-protected metal clusters. *J. Phys. Chem. C*, 119:28666–28678, 2015. DOI: 10.1021/acs.jpcc.5b10235.

[168] Li, Y., Higaki, T., Du, X., and Jin, R. Chirality and surface bonding correlation in atomically precise metal nanoclusters. *Adv. Mater.*, 1905488, 2020. DOI: 10.1002/adma.201905488. 57

[169] Kumar, S. and Jin, R. Water-soluble $Au_{25}(Capt)_{18}$ nanoclusters: Synthesis, thermal stability, and optical properties. *Nanoscale*, 4:4222–4227, 2012. DOI: 10.1039/c2nr30833a. 57, 58

[170] Dolamic, I., Varnholt, B., and Bürgi, T. Chirality transfer from gold nanocluster to adsorbate evidenced by vibrational circular dichroism. *Nat. Commun.*, 6:7117, 2015. DOI: 10.1038/ncomms8117. 57

[171] Dolamic, I., Knoppe, S., Dass, A., and Bürgi, T. First enantioseparation and circular dichroism spectra of Au_{38} clusters protected by achiral ligands. *Nat. Commun.*, 3:798, 2012. DOI: 10.1038/ncomms1802. 57

[172] Knoppe, S., Dolamic, I., Dass, A., and Bürgi, T. Separation of enantiomers and CD spectra of $Au_{40}(SCH_2CH_2Ph)_{24}$: Spectroscopic evidence for intrinsic chirality. *Angew. Chem. Int. Ed.*, 51:7589–7591, 2012. DOI: 10.1002/anie.201202369. 57

[173] Murray, R. W. Nanoelectrochemistry: Metal nanoparticles, nanoelectrodes, and nanopores. *Chem. Rev.*, 108:2688–2720, 2008. DOI: 10.1021/cr068077e. 57

[174] Kwak, K. and Lee, D. Electrochemistry of atomically precise metal nanoclusters. *Acc. Chem. Res.*, 52:12–22, 2019. DOI: 10.1021/acs.accounts.8b00379. 57

[175] Lee, D., Donkers, R. L., Wang, G., Harper, A. S., and Murray, R. W. Electrochemistry and optical absorbance and luminescence of molecule-like Au_{38} nanoparticles. *J. Am. Chem. Soc.*, 126:6193–6199, 2004. DOI: 10.1021/ja049605b. 57, 58

[176] Kwak, K. and Lee, D. Electrochemical characterization of water-soluble Au_{25} nanoclusters enabled by phase-transfer reaction. *J. Phys. Chem. Lett.*, 3:2476–2481, 2012. DOI: 10.1021/jz301059w. 58

[177] Wu, Z. Anti-galvanic reduction of thiolate-protected gold and silver nanoparticles. *Angew. Chem. Int. Ed.*, 51:2934–2938, 2012. DOI: 10.1002/anie.201107822. 59

[178] Wang, M., Wu, Z., Chu, Z., Yang, J., and Yao, C. Chemico-physical synthesis of surfactant-and ligand-free gold nanoparticles and their anti-galvanic reduction property. *Chem. Asian J.*, 9:1006–1010, 2014. DOI: 10.1002/asia.201301562. 59

[179] Sun, J., Wu, H., and Jin, Y. Synthesis of thiolated Ag/Au bimetallic nanoclusters exhibiting an anti-galvanic reduction mechanism and composition-dependent fluorescence. *Nanoscale*, 6:5449–5457, 2014. DOI: 10.1039/c4nr00445k. 59

[180] Zhang, W., Zhuang, S., Liao, L., Dong, H., Xia, N., Li, J., Deng, H., and Wu, Z. Two-way alloying and dealloying of cadmium in metalloid gold clusters. *Inorg. Chem.*, 58:5388–5392, 2019. DOI: 10.1021/acs.inorgchem.9b00125. 59

[181] Tuboltsev, V., Savin, A., Pirojenko, A., and Räisänen, J. Magnetism in nanocrystalline gold. *ACS Nano*, 7:6691–6699, 2013. DOI: 10.1021/nn401914b. 60

[182] Zhu, M., Aikens, C. M., Hendrich, M. P., Gupta, R., Qian, H., Schatz, G. C., and Jin, R. Reversible switching of magnetism in thiolate-protected Au_{25} superatoms. *J. Am. Chem. Soc.*, 131:2490–2492, 2009. DOI: 10.1021/ja809157f. 60

[183] Wu, Z., Chen, J., and Jin, R. One-pot synthesis of $Au_{25}(SG)_{18}$ 2- and 4-nm gold nanoparticles and comparison of their size-dependent properties. *Adv. Funct. Mater.*, 21:177–183, 2011. DOI: 10.1002/adfm.201001120. 60

[184] Krishna, K. S., Tarakeshwar, P., Mujica, V., and Kumar, C. S. S. R. Chemically induced magnetism in atomically precise gold clusters. *Small*, 10:907–911, 2014. DOI: 10.1002/smll.201302393. 60

[185] Negishi, Y., Tsunoyama, H., Suzuki, M., Kawamura, N., Matsushita, M. M., Maruyama, K., Sugawara, T., Yokoyama, T., and Tsukuda, T. X-ray magnetic circular dichroism of size-selected, thiolated gold clusters. *J. Am. Chem. Soc.*, 128:12034–12035, 2006. DOI: 10.1021/ja062815z. 61

[186] Agrachev, M., Antonello, S., Dainese, T., Gascón, J. A., Pan, F., Rissanen, K., Ruzzi, M., Venzo, A., Zoleo, A., and Maran, F. A magnetic look into the protecting layer of Au_{25} clusters. *Chem. Sci.*, 7:6910–6918, 2016. DOI: 10.1039/c6sc03691k. 61

CHAPTER 4

Applications of Atomically Precise Metal Nanoclusters

4.1 INTRODUCTION

Metal NCs (NCs) [1–5] have recently attracted a great deal of interest in catalysis [6–8], sensing [9–11], biology, and biomedicine [12–20], as well as energy [2, 12, 21] applications due to many unique properties of NCs, such as the well-defined structures, optical absorption, luminescence, catalytic activity, and magnetism, to name a few. Such properties are not seen in the corresponding bulk materials or nanomaterials with larger sizes [2, 12]. More significantly, the properties of metal NCs are sensitive to the size [22], composition [23, 24], surface [7, 25], charge state [26], and structure [27], therefore providing opportunities for tailoring of properties and guiding the design for specific application of metal NCs.

One of the most attractive properties of metal NCs is their strong photoluminescence, combined with good photostability, large Stokes shift, and high emission [11, 16, 19]. Each size of NCs has its feature profile of optical absorption, which is different from plasmonic NPs. These unique aspects make metal NCs highly promising in chemical sensing and biosensing applications. For further engineering toward the biomedical applications, the ligand shell in the outer layer of the metal NCs can be tailored by employing biomolecules, such as polymers, dendrimers, proteins, peptides, and DNAs to control the biocompatibility and enhance the luminescence intensity [14, 19], thus establishing metal NCs as a new class of ultrasmall biocompatible fluorophores in bioassays of DNA/protein, biolabeling, bioimaging, drug delivery, biotherapy, etc.

Metal NCs have also shown great potential for catalyzing many important chemical transformation reactions owing to their good activity (higher surface-to-volume ratio than conventional NPs) and high selectivity under relatively mild conditions [2]. More importantly, the active sites and catalytic mechanism could be analyzed based on the precise structures of metal NCs. To elucidate the structure-property relationships, the catalytic performance has been tuned by means of the well-controlled size, composition, and structure of metal NCs. Ligands are crucial for the preparation of metal NCs as stabilizers, but the NCs capped by ligands often show lower activity than partially or completely exposed ones. Therefore, (mild) thermal treatments are employed to remove the undesirable ligands without causing much change to the particle size. On the other hand, the ligands can also play important roles in catalytic reactions and improve the selectivity in some cases. Overall, metal NCs have emerged as a new class of model catalysts in many reactions [2, 5–8, 28], such as the oxidation, hydrogenation, C-C coupling, electron-transfer catalysis, three-component coupling reaction, photocatalysis, electrocatalysis, etc.

In this chapter, the promising applications of metal NCs in catalysis, sensing, biology, biomedicine, and energy are summarized, followed by a short perspective for future work.

4.2   CATALYSIS

Among the metal NCs, Au NCs are more prevalent in the investigation of catalytic applications, while the reports on the catalysis of Ag/Cu NCs are less partly due to the immature synthesis and instability.

In terms of catalysis, gold was once considered to be catalytically inactive for a long time before the work of Haruta's group in the late 1980s [29]. Since then, much work has explored the high activity of small Au NPs in various reactions. However, the conventional Au nanocatalysts are polydispersed in size and structure, resulting in the lack of understanding the structure-property relationships [28]. Recent advances in the atomically precise synthesis of NCs not only promote the application of Au catalysts but also enable the elucidation of the active sites and structure-related performance at the atomic level [20].

4.2.1   OXIDATION

4.2.1.1   CO Oxidation

The CO oxidation process is of importance in practical processes, such as the removal of CO impurity from H_2 in fuel cells [30]. The conventional nanogold catalysts exhibit excellent activity for CO oxidation at mild or even low temperatures (e.g., $-70°C$) [29]. However, $Au_{25}(SR)_{18}$ (SR: thiolate) has no notable catalytic activity (even at 200°C) when loaded on active support such as TiO_2 [31]. Nie et al. found that the pretreatment of $Au_{25}(SR)_{18}$/metal oxide catalysts significantly affected the catalytic activity during CO oxidation (Fig. 4.1). The catalytic activity was highly improved after O_2 pretreatment of the catalyst at 150°C for 1.5 h. Since the pretreatment temperature is well below the thiolate desorption temperature ($\sim 200°C$), Au_{25} NCs should be intact during the pretreatment without losing ligands. No further enhancement effect is observed when changing the pretreatment atmosphere from O_2 to N_2 or increasing the O_2 pretreatment temperature to 250°C, indicating that the removal of surface thiolates is not always necessary. It is proposed that the interface of $Au_{25}(SR)_{18}$/metal oxide is critical for the catalytic reaction, and O_2 is converted to active oxygen species (e.g., hydroperoxide) at the interface during the pretreatment. They also investigated the CO oxidation of Au_{25} supported with different metal oxides (including TiO_2, CeO_2, and Fe_2O_3), and found that $Au_{25}(SR)_{18}$/CeO_2 catalyst was much more active than the others.

Although the trend of nanogold-catalyzed CO oxidation has been reported [32], the precise size dependence was not clear due to their inherent polydispersity. With atomically precise NCs, the CO oxidation was investigated on Au_{25}, Au_{38}, Au_{144}, and Au_{333} protected by the

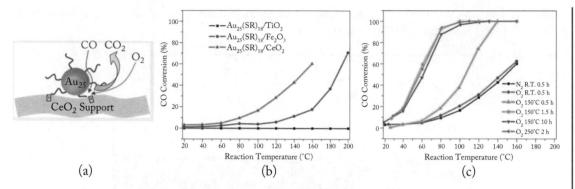

(a)   (b)   (c)

Figure 4.1: (a) Proposed model for CO oxidation at the perimeter sites of $Au_{25}(SR)_{18}/CeO_2$ catalyst. (b) Reaction temperature dependence of CO conversion over different catalysts. (c) Reaction temperature dependence of CO conversion over $Au_{25}(SR)_{18}/CeO_2$ catalyst after different pretreatments. Reproduced with permission from [31]. Copyright 2012 American Chemical Society.

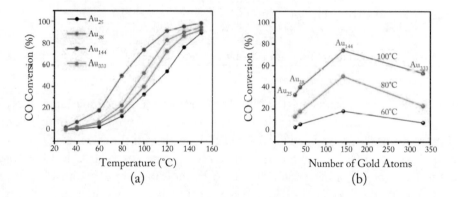

(a)   (b)

Figure 4.2: (a) The light-off curves of different sized NCs. (b) A volcano-like trend of size effect in CO oxidation. Reproduced with permission from [22]. Copyright 2016 Rongchao Jin.

same thiolate (PET) [22]. A volcano-like trend of size effect was found and the optimum size was Au_{144} among the four sizes, which lay in the metallic to molecular-state transition regime, indicating that the high activity are not fully originated from quantum effect (see Fig. 4.2).

4.2.1.2   Styrene Oxidation

The styrene oxidation reaction is an important organic transformation reaction for the value-added fine chemical—styrene oxide [33]. In 2008, Turner et al. found Au_{55} showed tremendous catalytic activity in the selective oxidation of styrene by O_2 while conventional supported gold

Table 4.1: The catalytic performance of thiolate capped $Au_n(SR)_m$ cluster catalysts for selective oxidation of styrene with O_2. Data from [35].

Catalysts	Conversion (%)	Selectivity (%)		
		1	2	3
1. $Au_{25}(SC_2H_4Ph)_{18}$	27 ± 1.0	70	24	6
2. $Au_{25}(SC_6H_{13})_{18}$	25 ± 0.8	69	26	5
3. $Au_{38}(SC_2H_4Ph)_{24}$	14 ± 1.2	72	24	4
4. $Au_{38}(SC_6H_{13})_{24}$	15 ± 1.2	71	25	4
5. $Au_{144}(SC_2H_4Ph)_{60}$	12 ± 0.5	80	20	Trace
6. $Au_{144}(SC_{12}H_{25})_{60}$	11 ± 0.8	84	16	Trace
7. ~ 3 nm Au NPs[a]	6 ± 1.2	82	18	Trace
[a] These particles are stabilized by octanethiolate				

NPs larger than 2 nm were inactive [34]. Although the structural determination of Au_{55} has not been successful yet thus far, the distinct electronic structure of Au NCs apparently contributes large to the extraordinary catalytic activity. Zhu et al. investigated the styrene oxidation of three sizes of Au NCs (including Au_{25}, Au_{38}, and Au_{144}) in toluene at ~ 100°C in O_2 (1 bar) [35]. The product was composed of benzaldehyde (70–80%, mol%), styrene oxide (15–25%, mol%), and acetophenone (0–5%, mol%), and Au_{25} was slightly more active than the other two NCs (Table 4.1). They also found the highest activity of the three NCs when using tert-butyl hydroperoxide (TBHP) as the oxidant.

Liu et al. reported very high conversion of styrene (100%) and yield of styrene epoxide (92%) using water-soluble $Au_{25}(SG)_{18}$/HAP (SG: glutathionate, HAP: hydroxyapatite) as a catalyst (Fig. 4.3) [36]. The supported catalyst was calcinated at 300°C for removal of the thiolate ligands before use. Recently, Wang et al. found alloy NCs ($Ag_{46}Au_{24}$ and $Ag_{32}Au_{12}$) showed different activity and selectivity compared with homogold or homosilver NCs in styrene oxidation reaction [37].

4.2.1.3   Alcohol Oxidation

Nanogold catalyst (e.g., poly(N-vinyl-2-pyrrolidone) (PVP) protected Au NCs) is also efficient for promoting the selective oxidation of alcohols [38]. Yoskamtorn et al. prepared hierarchically

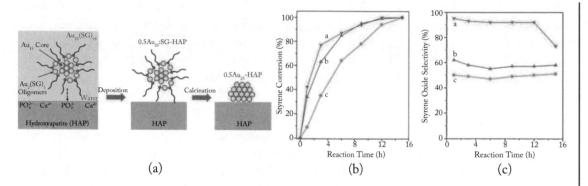

(a)   (b)   (c)

Figure 4.3: (a) Synthetic procedure of HAP-supported Au$_{25}$. Time evolution of (b) conversion of styrene and (c) selectivity of styrene epoxide for (a) 0.5 Au$_{25}$-HAP, (b) 0.5Au-HAP (adsorption), and (c) 0.5Au-HAP(impregnation). Reproduced with permission from [36]. Copyright 2010 Royal Society of Chemistry.

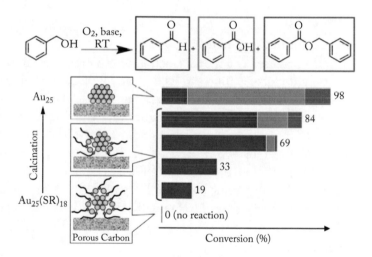

Figure 4.4: Catalytic performances of Au$_{25}$/HPCS for aerobic oxidation of benzyl alcohol. Reproduced with permission from [39]. Copyright 2014 American Chemical Society.

porous carbon nanosheets (HPCS)-supported Au$_{25}$ NC catalysts by calcination at 400–500°C for 2–4 h (Fig. 4.4) [39]. With the number of residual thiolates on Au$_{25}$ NCs increasing, the selectivity for benzaldehyde formation was remarkably improved while the conversion of benzyl alcohol was reduced [39].

Wang et al. reported the catalytic activity of the quasi-isomeric Au$_{28}$(SR)$_{20}$ (SR = tertbutyl benzenethiolate, TBBT, or cyclohexanethiolate (CHT) for the oxidation of benzyl alcohol [40]. Except for the ligand difference, the two NCs have also different types of Au-S staples: four dimeric staples for $Au_{28}(TBBT)_{20}$, two monomeric staples and two trimeric staples for $Au_{28}(CHT)_{20}$. The supported $Au_{28}(TBBT)_{20}$ always exhibited higher conversion of benzyl alcohol than $Au_{28}(CHT)_{20}$ when supported on different metal oxides, indicating that the dimeric staples might be the more active site for alcohol oxidation.

4.2.1.4   Cyclohexane Oxidation

Oxidation of cyclohexane can produce important intermediates in the chemical industry, such as cyclohexanone or cyclohexanol. Liu et al. studied the catalysis by $Au_{10}(SG)_{10}$, $Au_{18}(SG)_{14}$, $Au_{25}(SG)_{18}$, and $Au_{39}(SG)_{24}$ on HAP for cyclohexane oxidation using O_2 and TBHP at 150°C [41]. The total selectivity reached $\sim$ 99% with $\sim$ 15% cyclohexane conversion; Table 4.2.

The $Au_{28}(TBBT)_{20}$ and $Au_{28}(CHT)_{20}$ NCs have also been used to catalyzed cyclohexane oxidation, and $Au_{28}(TBBT)_{20}$ was found to show better activity than $Au_{28}(CHT)_{20}$ [40]. Moreover, the two Au_{28} NCs exhibited a higher selectivity toward cyclohexanol than cyclohexanone.

4.2.2   HYDROGENATION

4.2.2.1   Hydrogenation of Nitro Compounds

4-nitrophenol reduction is often used as a model reaction to explore the catalytic application of nanomaterials. The reaction produces aniline and its derivatives. Tian et al. discovered the first pair of structural isomer NCs protected by PET, Au_{38T} and Au_{38Q} (PET: phenylethanethiolate), and revealed that Au_{38T} exhibits remarkably higher catalytic activity (44%) than Au_{38Q} (0%) at low temperature (0°C) in 4-nitrophenol reduction with $NaBH_4$, indicating the structure sensitivity for NC catalysis (Table 4.3) [42]. Li et al. synthesized a new $Au_{44}(PET)_{32}$ by reacting Au_{25} with $Cu(NO_3)_2$ and found that $Au_{44}(PET)_{32}$ exhibits the highest catalytic reduction activity of nitrophenol among some common NCs [43]. This work demonstrated that slight difference in composition may also lead to dramatic changes in catalysis performance.

4.2.2.2   Semihydrogenation of Alkynes

The semihydrogenation of alkynes produces most of the olefins for polymers. Li et al. studied the catalytic performance of the spherical $Au_{25}(PET)_{18}$ and rod-shaped $Au_{25}(PPh_3)_{10}(C \equiv CPh)_5X_2$ (X = Br, Cl) NCs supported on oxides for semihydrogenation of terminal alkynes to alkenes [44]. The conversion of terminal alkynes reached $\sim$ 100% and the selectivity for alkene products ($\sim$ 100%) was also excellent; Table 4.4. Of note, the semihydrogenation of internal alkynes requires ligand-off catalysts.

Recently, Wan et al. investigated the catalytic semihydrogenation of Au_{38} NCs with FCC kernel. They found that alkynyl-protected Au_{38} are much more active (> 97%) than the thiolated Au_{38} (< 2%), indicating a huge ligand effect in catalysis [45]. Besides, Chen et al. investigated the hydrogenation of acetylene using encapsulated Au_{25} in microporous silica as catalyst. This

Table 4.2: Oxidation of cyclohexane using differently sized Au NCs. Data from [41].

Catalysts	Conversion (%)	Selectivity (%)		TOF ($\times 10^4 \, h^{-1}$)[a]
		2	3	
1. Au$_{10}$/HAP	11.6	58	41	1.19
2. Au$_{18}$/HAP	12.9	49	50	1.33
3. Au$_{25}$/HAP	14.2	50	49	1.37
4. Au$_{39}$/HAP	14.9	50	49	1.85
5. Au$_{\sim 85}$/HAP	6.7	53	41	0.68
6. Au$_{25}$/HAP$_{second}$	13.9	51	47	
7. Au$_{25}$/HAP$_{third}$	13.7	50	48	

[a] TOF was calculated from the amount of 1 recovered for the following reaction conditions and materials: Au$_n$/HAP (100 mg), TBHP (10 mg), 1 (15 mL), and O_2 (1 MPa, ~100 mL) at 150°C for 30 min.

Table 4.3: Catalytic activities of Au$_{25}$, Au$_{38T}$, and Au$_{38Q}$ in 4-nitrophenol reduction. Data from [43].

Catalysts, (0.1 mol%)	Yield (%)
Au$_{25}$	0
Au$_{38Q}$	0
Au$_{38T}$	44

Table 4.4: Comparison of conversion in semihydrogenation catalyzed by ligand-on and ligand-off catalysts. Data from [44].

Entry	Catalysts	R^1	R^2	Conversion	Selectivity
1	Ligand-on Au_{25} sphere	PhC_2H_4	H	>99	
2		Ph	CH_3	<1	
3		Ph	Ph	<1	
4		n-C_6H_{13}	CO_2CH_3	<1	
5	Ligand-off Au_{25}	PhC_2H_4	H	95.6	
6		Ph	CH_3	52.8	97
7		Ph	Ph	59.7	>99
8		n-C_6H_{13}	CO_2CH_3	52.6	99
9	Ligand-on Au_{25} rod	PhC_2H_4	H	99.8	
10		Ph	CH_3	<1	
11		Ph	Ph	<1	
12		n-C_6H_{13}	CO_2CH_3	<1	

Reaction scheme: $R^1 \!\!-\!\!\equiv\!\!-\!\! R^2$ with Au_{25}/TiO_2, Pyridine, EtOH/H_2O, 100°C, 20 bar H_2, 20h, yielding the cis-alkene product.

protocol avoided the aggregation of NCs during thermal pretreatment at 500°C. The reaction was performed at ambient pressure and the acetylene conversion efficiency reached 95% at 280°C with a selectivity of 98% toward ethylene [46].

4.2.2.3 Selective Hydrogenation

In 2010, Zhu et al. investigated the selective hydrogenation of α,β-unsaturated ketones using supported Au_{25} NCs as catalyst [47]. They found that the chemoselectivity of unsaturated alcohol in the hydrogenation of benzalacetone was nearly 100% even at low temperatures (e.g., 0°C). The core-shell structure and unique electronic properties of Au_{25} were proposed to be responsible for the catalytic performance; Fig. 4.5. The electron-rich Au_{13} core favors adsorption of $C = O$ and the electron-deficient Au_{12} shell is thought to provide sites for H_2 adsorption and dissociation, then nucleophilic hydrogen attacks the activated $C = O$ group, leading to the unsaturated alcohol product. Following that, the first stereoselective hydrogenation of bicyclic ketone catalyzed by Au NCs was also reported by them [48]. The hydrogenation reaction of 7-(phenylmethyl)-3-oxa-7-azabicyclo[3.3.1] nonan-9-one was conducted at room temperature under a H_2 flow and the stereoselectivity of exo-alcohol was nearly 100%.

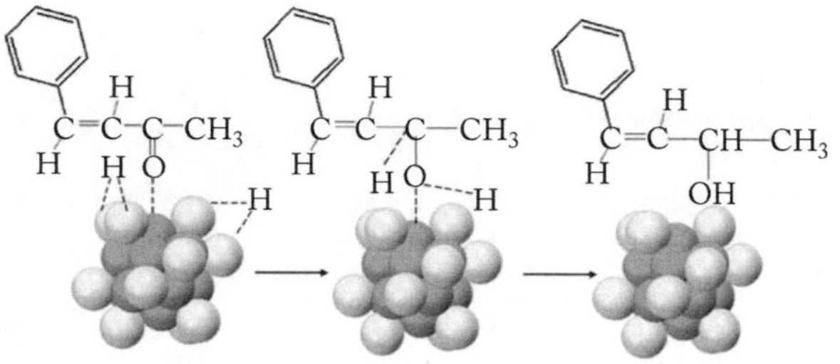

Figure 4.5: The proposed mechanism of the chemoselective hydrogenation of α,β-unsaturated ketone to unsaturated alcohol catalyzed by $Au_{25}(SR)_{18}$. Reproduced with permission from [47]. Copyright 2010 Wiley VCH.

Li et al. reported the formation of α,β-unsaturated ketones and aldehydes from propargylic acetates in high isolated yields catalyzed by negative Au_{25} [49]. For example, the conversion of 1,3-diphenylprop-2-ynyl acetate reached 80% using Au_{25} and K_2CO_3 in DMSO (Table 4.5). The reaction was found solvent sensitive and the catalyst is recoverable and recyclable.

4.2.3 C-C COUPLING REACTION

The C–C coupling reactions are very useful for the synthesis of natural products, pharmaceuticals, polymers, etc [28]. Au NCs have been proved well active to catalyze C–C coupling reaction. Both Jin and Li have already reviewed the applications of Au NCs in C–C coupling reactions including the Ullmann-type homocoupling, Sonogashira cross-coupling, and Suzuki coupling [50, 51]. In this subsection, we just discuss several works for reason of a simpler presentation. In 2012, Li et al. investigated the catalytic activity of Au_{25} for the carbon—carbon homocoupling reaction of iodobenzene to the biphenyl product [52]. The supported Au_{25} on CeO_2 catalyst showed highest activity (up to 99.8% conversion of iodobenzene). Later, the Ullmann heterocoupling reaction was examined by Au_{25} NCs with different thiolate ligands (e.g., naphthalenethiolate, benzenethiolate, hexanethiolate, and 2-phenylethanethiolate) [53]. The aromatic thiolated Au_{25} NCs were revealed to have better catalytic performance compared with the nonaromatic thiolated Au NCs in both the conversion and selectivity.

The Suzuki cross-coupling reaction of phenylboronic acid and p-iodoanisole was efficiently catalyzed by TiO_2-supported Au_{25} NCs with the addition of imidazolium-based ionic liquids (up to 99% conversion) [54]. The acidic proton of the imidazolium ions was revealed to be a key role for the detachment of the ligands or Au atoms from the NC surface to afford active sites for catalytic reactions.

Li et al. also studied Sonogashira cross-coupling reaction between phenylacetylene and p-iodoanisole catalyzed by Au_{25} NCs supported on oxides (such as CeO_2, TiO_2, MgO, and SiO_2). The reaction achieved with high conversion (up 96.1%) and excellent selectivity (88.1%) [55].

Table 4.5: The isolated yields of (E)-chalcone under various conditions. Data from [49].

Entry	Catalysts	Additive	Temp (°C)	Yield
1	$Au_{25}(SR)_{18}^-$	None	80	0
2	None	K_2CO_3	80	0
3	$Au_{25}(SR)_{18}^-$	K_2CO_3	80	76
4	$Au_{25}(SR)_{18}^-$	K_2CO_3	25	0
5	$Au_{25}(SR)_{18}^-$	K_2CO_3	60	62
6	$Au_{25}(SR)_{18}^-$	KOH	80	40
7	$Au_{25}(SR)_{18}^-$	CH_3COOH	80	Trace
8	$Au_{25}(SR)_{18}^0$	K_2CO_3	80	75
9	$Au_{38}(SR)_{24}$	K_2CO_3	80	Trace
10	$Au_{144}(SR)_{60}$	K_2CO_3	80	Trace
11	~ 3 nm Au NP[a]	K_2CO_3	80	Trace
12	~ 3 nm Au NP[b]	K_2CO_3	80	Trace
13	~ 23 nm Au NP	K_2CO_3	80	0
14	$Au_{25}(SR)_{18}^-$	K_2CO_3	80	48[c]
15[c]	$Au_{25}(SR)_{18}^-$	K_2CO_3	80	48
16[c]	$Au_{25}(SR)_{18}^-$	K_2CO_3	80	Trace[d]
17	$Au_{25}(SR)_{18}^-$	K_2CO_3	100	78
18	$Au_{25}(SR)_{18}^-$	K_2CO_3	100	78[e]
19	$Au_{25}(SR)_{18}^-$	K_2CO_3	100	80[f]
20	$Au_{25}(SR)_{18}^-$	K_2CO_3	100	84[f,g]

[a] Au NP was protected by phenylethanethiolate. [b] Au NP was not protected by ligands. [c] DMF was used as solvent. [d] Water was not added. [e] Reaction time was prolonged to 5 h. [f] Solvent: 0.5 mL DMF and 0.05 mL H_2O. [g] 10.0 mg catalyst was added.

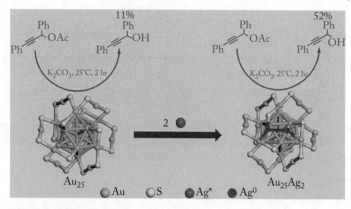

Figure 4.6: Catalytic performance of Au_{25} and $Au_{25}Ag_2$ for the Hydrolysis of 1,3-Diphenylprop-2-ynyl Acetate. Reproduced with permission from [59]. Copyright 2015 American Chemical Society.

DFT calculations demonstrated that the strong co-adsorption of phenylacetylene and iodobenzene on the surface of the Au_{25} facilitates configuration ready for coupling.

4.2.4   THREE-COMPONENT COUPLING REACTION

Three component coupling of alkynes, aldehydes, and amines (A^3-coupling) is an important reaction in organic chemistry to prepare propargylamines and their derivatives. Both $Au_{25}(PET)_{18}$ and $Au_{38}(PET)_{24}$ show efficient activity, and it is worth noting only a very low loading (0.01 mol%) is needed for Au_{38} catalyzed A^3-coupling [56, 57]. Besides, Li et al. prepared a novel Cd-doped gold NC $Au_{26}Cd_5$ by doping and peeling of Au_{25} [58]. The Cd-doped Au NCs were found to catalyze the A^3 reaction with a very high efficiency. The cooperation of the exerted cadmium atoms and the neighbor gold atoms may be responsible for the catalytic activity.

4.2.5   HYDROLYSIS

Yao et al. synthesized a novel alloy $Au_{25}Ag_2(PET)_{18}$ NCs by depositing two Ag atoms on Au_{25} surface [59]. They found the hydrolysis of 1,3-diphenylprop-2-ynyl was remarkably accelerated using $Au_{25}Ag_2$ as catalyst with a 52% conversion, implying that the two Ag atoms on Au_{25} provide more active sites for the reaction.

4.2.6   PHOTOCATALYSIS

Au NPs often exhibit localized surface plasmon resonance under light irradiation. However, drastic changes occur to the optical properties when particles size fall in the region of NCs (e.g., 0–3 nm), which evoke the exploration of their photocatalysis. In 2010, Kogo et al. reported the

Table 4.6: Comparison of catalytic performances of the $[Au_{25}(PET)_{18}]^-$, $Au_{38}S_2(SAdm)_{20}$, and $Au_{38}(PET)_{24}$ NCs in the selective oxidation of methyl phenyl sulfide. Data from [62].

Entry	Au Nanocluster	Light (nm)	Conversion (°C)	Selectivity (%)
1[a]	$Au_{38}S_2(Adm)_{20}$	0	nr	0
2	$Au_{25}(PET)_{18}^-$	532	18	100
3	$Au_{38}S_2(Adm)_{20}$	532	57	100
4[b]	$Au_{38}S_2(Adm)_{20}$	532	58	100
5[c]	$Au_{38}S_2(Adm)_{20}$	532	56	100
6[d]		532	nr	
7[e]	$Au_{38}S_2(Adm)_{20}$	532	nr	
8	$Au_{38}(SR)_{24}$	532	Trace	

[a] The catalytic reaction was carried out in the absence of light. [b] Second reuse of the $Au_{38}S_2(SAdm)_{20}$ recovered from entry 3. [c] Third reuse of the $Au_{38}S_2(SAdm)_{20}$ recovered from entry 3. [d] No gold nanoclusters were added to the reaction system. [e] The catalytic reaction was carried out under a N_2 atmosphere.

photocatalysis of Au_{25}-modified TiO_2 for the oxidation of phenol derivatives and ferrocyanide and reduction of Ag^+, Cu^{2+}, and oxygen under visible and NIR light (≤ 860 nm) [60]. Yu et al. investigated the visible-light photocatalytic properties of $Au_{25}(SR)_{18}/TiO_2$ composites for the degradation of methyl orange. Au_{25} NCs loaded on the TiO_2 surface were thought as a small-band-gap semiconductor to absorb visible light, then producing electron-hole separation and singlet oxygen for the reaction [61].

Recently, Li et al. found $Au_{38}S_2(SAdm)_{20}$ (SAdm: 1-adamantanethiolate) NCs was highly photosensitive to the generation of singlet oxygen under visible/near-IR (e.g., 532 and 650 nm) irradiation [62]. In this work, Au_{38Q} gave rise to low catalytic conversion of methyl phenyl sulfide, while $Au_{38}S_2(Adm)_{20}$ showed 57% conversion with 100% selectivity under the same reaction conditions, see Table 4.6.

Chen et al. reported the good photocatalytic properties of $Au_{25}(PPh_3)_{10}Cl_2(SC_3H_6SiO_3)_5/TiO_2$ catalyst for the selective oxidation of amines to imines under visible light [63]. Partial ligand removal under the reaction conditions provides active sites for the oxidation. The photocatalysis proceeded by a carbocation intermediate

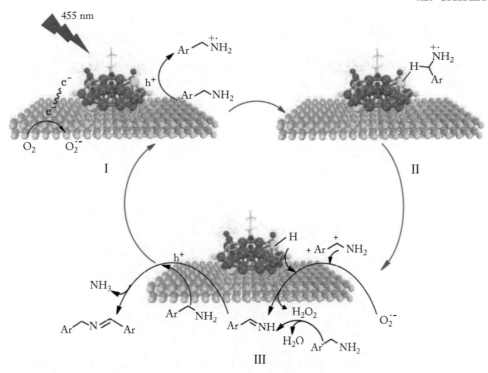

Figure 4.7: Proposed mechanism for photo-oxidation of benzylamine catalyzed by rod Au_{25}/TiO_2. Reproduced with permission from [63]. Copyright 2017 American Chemical Society.

and an Au–H intermediate species, which was confirmed by a trapping agent, as shown in Fig. 4.7 [63]. Liu et al. investigated the photocatalytic degradation of organic pollutants using $[Au_{23-x}Ag_x(Adm)_{15}]$ ($x = 4 \sim 7$) NC as catalyst [64]. Due to the doping of Ag atoms, $Au_{23-x}Ag_x/TiO_2$ shows much better catalytic activities than Au_{23}/TiO_2 in the photocatalytic degradation of RhB and phenol under visible-light irradiation.

4.2.7 ELECTROCATALYSIS

Compared to the poor catalytic performance of bulk gold, the emerging Au NCs show great potential in electrocatalysis with enhanced activity and controllable properties (such as size, surface, structure and compositions) [65]. In 2009, Chen et al. studied the size effect of gold NCs (including Au_{11}, Au_{25}, Au_{55}, and Au_{140}) in the oxygen electroreduction [66]. They found smaller NCs exhibited much higher electrocatalytic activity than large ones.

Kauffman et al. reported the excellent catalytic activity of supported Au_{25} for the electrochemical conversion of CO_2 into CO [67]. Peak CO_2 conversion occurred at -1 V (vs RHE)

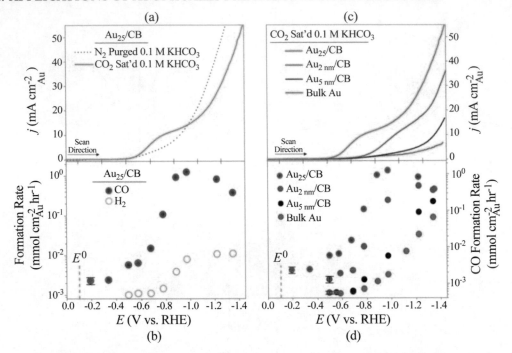

Figure 4.8: (a) Linear sweep voltammograms (LSV) of carbon black (CB) supported Au_{25}. (b) Potential-dependent H_2 and CO formation rates for Au_{25}/CB. (c) LSVs of various Au catalysts. (d) Potential-dependent CO formation rates for the various Au catalysts. Reproduced with permission from [67]. Copyright 2012 American Chemical Society.

with approximately 100% efficiency and a rate 7–700 times higher than that for larger Au NPs. The charge states of Au_{25}^q ($q = -1, 0, +1$) impact electrocatalytic activities by stabilizing reactant or product adsorption. Recently, Zhao et al. compared the electrocatalysis of spherical Au_{25} nanosphere and Au_{25} nanorod both supported on carbon black for electrochemical CO_2 reduction [68]. Spherical Au_{25} exhibits higher Faradaic efficiency for CO with higher formation rates than rod-shaped Au_{25}. The difference in catalytic performance was ascribed to the negative charge state and easily exposed active sites of the Au_{25} nanosphere.

Lu et al. investigated the electrocatalytic activity of $Au_{25}(SR)_{18}$ and $Au_{24}Pt(SR)_{18}$ (SR: 1-dodecanthiol) [69]. They found single-Pt doping of Au_{25} efficiently enhance the electrocatalytic activity for the direct formic acid oxidation (FAO) to carbon dioxide. The mass activity of Pt_1Au_{24} (3.7 A mg_{Pt+Au}^{-1}) is nearly 12 and 34 times greater than that of Pt NCs and the commercial Pt/C catalyst, respectively. This work revealed the structure-activity relationship and provided a strategy to tune the catalytic properties of nanomaterials at the atomic level. Very recently, Zhuang et al. reported a novel bimetallic $Au_{47}Cd_2(TBBT)_{31}$ NCs by a two-phase anti-galvanic reduction method using $Au_{44}(TBBT)_{28}$ as precursors. Such bimetal NCs not only show a unique hard-sphere random close-packed kernel structure but also have higher Faradaic efficiencies for electrocatalytically reducing CO_2 to CO (96% at -0.57 V) than $Au_{44}(TBBT)_{28}$ and Au NPs [70].

4.3   SENSING

Sensing or detection of chemical and biological agents plays a fundamental role in environmental, biomedical, and forensic field [71]. Fortunately, metal NCs provide excellent scaffolds for sensing various compounds, due to their unique optoelectronic properties, high surface-to-volume ratio, controllable properties (such as size, surface, structure and compositions). Moreover, the atomically precise NCs are better for understanding the sensing mechanism including, for example, metallophilic interactions, electron or energy transfer, Förster resonance energy transfer (FRET), anti-galvanic reaction, etc. In an early stage, lots of fluorescent NCs templated by carboxylic acids, polymers, DNA, peptides, and proteins show good sensing sensitivity and selectivity but without well determination of atomic composition and structure [72–91]. Here we mainly focus on the sensing application of metal NCs with precise composition and structure. Table 4.7 lists some of the atomically precise metal NC-based sensors and their properties [92–103].

4.3.1   METAL IONS

Muhammed et al. report of the synthesis of Au_{23} NCs by the etching of $Au_{25}(SG)_{18}$ and found the use of Au_{23} for Cu^{2+} sensing through the fluorescence quenching [95]. In 2012, Wu et al. first reported the detection of Ag^+ based on the fluorescence enhancement of $Au_{25}(SG)_{18}$ [92], with the good selectivity among 20 types of metal cations (Fig. 4.9). The fluorescence enhancement of NCs rarely happens when interacting with metal ions since metal ions typically cause fluorescence quenching. In this work, the unique enhancement caused by Ag^+ was attributed to three factors: (i) the oxidation state change of Au_{25}, (ii) interaction of the neutral silver species with Au_{25}, and (iii) interaction of silver ion (Ag^+) with Au_{25}.

Xia et al. developed a facile strategy to synthesize Ag_{30} NCs protected by captopril and this NC shows excellent solubility in water and various polar organic solvents after protonation. They also demonstrated that Ag_{30} can be a potential colorimetric probe for Hg^{2+} with a detection limit of 6 ppb and well selectivity among 13 metal ions [93].

4.3.2   ANIONS

Wang et al. developed a novel fluorescent anion (iodide) sensor based on well-defined NCs [99]. They found $Au_{25}(SG)_{18}$ was a good candidate for iodide sensing with a detection limit of 400 nM. Enhanced fluorescence response of Au_{25} occurred by the addition of I^-, while the fluorescence intensity decreased induced by other 12 types of anions (F^-, Cl^-, Br^-, I^-, NO_3^-, ClO_4^-, HCO_3^-, IO_3^-, SO_4^{2-}, SO_3^{2-}, CH_3COO^-, $C_6H_5O_7^{3-}$) (Fig. 4.10).

Table 4.7: List of various NCs for sensing. Data from refs [92–100]. (*Continues.*)

Sensing Type	Nanocluster	Analyte	Selectivity	Detection Method	LOD	Media	Ref
Metal ions	$Au_{25}(SG)_{18}$	Ag^+	Pb^{2+}, Cd^{2+}, Hg^{2+}, Cu^{2+}, Zn^{2+}, Ni^{2+}, Co^{3+}, Tb^{3+}, Eu^{3+}, Pd^{2+}, Fe^{2+}, Fe^{3+}, Mg^{2+}, K^+, Na^+, Ca^{2+}, Cr^{3+}, Mn^{2+}, Au^{3+}	Flourescence enhancement	200 nM	Aqueous solution	92
	$Ag_{30}(Capt)_{18}$	Hg^{2+}	Na^+, K^+, Ca^{2+}, Mg^{2+}, Ba^{2+}, Cr^{3+}, Fe^{3+}, Fe^{2+}, Co^{2+}, Ni^{2+}, Zn^{2+}, Cd^{2+}, Pb^{2+}, Cu^{2+}	Colorimetric changes	6 ppb	Aqueous solution, lake water, soil solution	93
	$Ag_{62}S_{13}(SBu^t)_{32}$	Cu^{2+}	K^+, Na^+, Cd^{2+}, Zn^{2+}, Pb^{2+}, Ni^{2+}, Mn^{2+}, Mg^{2+}, Hg^{2+}, Cr^{3+}, Ba^{2+}, Fe^{3+}	Flourescence quenching	Not mentioned	$MeCN/H_2O$ 1/1	94
	Au_{23}	Cu^{2+}	Au^{3+}, Ag^+, Ni^{2+}, Ca^{2+}, Mg^{2+}, Na^+, Pb^{2+}, Hg^{2+}, Cd^{2+}	Flourescence quenching	Not mentioned	Aqueous solution	95
	$Au@SiO_2@Ag_{15}$	Hg^{2+}	Ni^{2+}, Cd^{2+}, Pb^{2+}, Cu^{2+}	Flourescence quenching	0.1 zeptomoles	Aqueous solution	96
	Au_{15}	Cu^{2+}	Hg^{2+}, As^{3+}, As^{5+}	Flourescence quenching	1 ppm	Aqueous solution	97
	BSA-protected Au_{25}	Hg^{2+}	–	Flourescence quenching	0.1 nM	Aqueous solution	98
Anions	$Au_{25}(SG)_{18}$	I^-	F^-, Cl^-, Br^-, NO_3^-, ClO_4^-, HCO_3^-, CO_3^{2-}, IO_3^-, SO_4^{2-}, SO_3^{2-}, CH_3COO^-, $C_6H_5O_7^{3-}$	Flourescence enhancement	400 nM	Aqueous solution	99
Small molecules	$Au@SiO_2@Ag_{15}$	2, 4, 6-	–	SERS and flourescence	10 ppb	Aqueous solution	96
	BSA-protected Au_{25}	Hydrogen	–	Flourescence quenching	10 nM	Aqueous solution	99
	$Ag_{29}(BDT)_{12}(PPh_3)_4$	Oxygen	Ar, N_2, CO_2	Flourescence quenching	Not mentioned	DCM solution	100

Table 4.7: (*Continued.*) List of various NCs for sensing. Data from refs [101–103].

Sensing Type	Nanocluster	Analyte	Selectivity	Detection Method	LOD	Media	Ref
Biomolecules	$Au_4(MPC)_4$	C-reactive protein	-	Flourescence quenching	5 nM	Aqueous solution	101
	$Cu_6(SG)_3$	Glucose	-	Electrochemical detection	4 μM	Aqueous solution	102
	Au_8	Hemoglobin	Myoglobin, cytochrome c, GSH, uric acid, glucose, Cl^-, NO_3^-, SO_4^{2-}, Na^+, K^+, Mg^{2+}, Ca^{2+}	Flourescence quenching	5 nM	Aqueous solution and blood sample	103

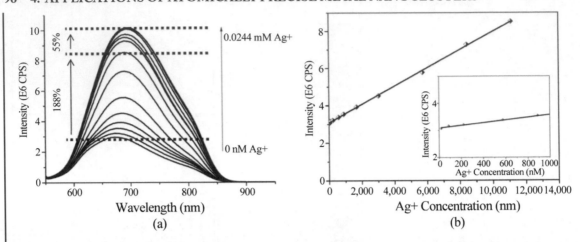

Figure 4.9: (a) Fluorescence spectra of $Au_{25}(SG)_{18}$ (~ 2.5 μM in H_2O) with successive titration of Ag^+. (b) Linear relation between the Ag^+ concentration (20.2 nM–11.1 μM) and the fluorescence intensity. Reproduced with permission from [92]. Copyright 2012 Wiley VCH.

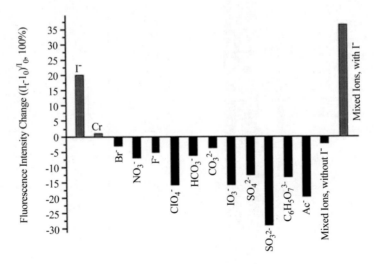

Figure 4.10: The influence of the different anions on the fluorescence of $Au_{25}(SG)_{18}$. Reproduced with permission from [99]. Copyright 2012 Royal Society of Chemistry.

4.3.3 SMALL MOLECULES

In 2012, a novel strategy for visual detection of 2,4,6-trinitrotoluene (TNT) at sub-zeptomole level was achieved using a hybrid material which are comprised of 15 Ag atoms embedded in bovine serum albumin on silica-coated Au mesoflowers, termed $Au@SiO_2@Ag_{15}$ [96]. Upon

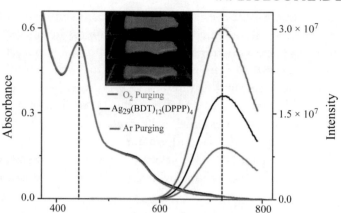

Figure 4.11: The optical absorption spectra and emission spectra of Ag_{29} after oxygenating and deoxygenating the cluster solution by purging O_2 (blue) and Ar (red). Reproduced with permission from [100]. Copyright 2018 Royal Society of Chemistry.

exposure to TNT, the fluorescence of the hybrids quenched and the Raman features from TNT (at 1209, 1361, 1535, 1619, and 2960 cm^{-1}) were detectable on the particle even at 100 ppt.

Recently, this group developed a prototypical O_2 sensor using the highly fluorescent $Ag_{29}(BDT)_{12}(PPh_3)_4$ NCs [100]. These clusters show fluorescence quenching and absorption maintaining in the presence of O_2. After purging by N_2, CO_2, or Ar and removing O_2 from the Ag_{29} solution, it can restore the fluorescence intensity.

4.3.4   BIOMOLECULES

Lu et al. reported a novel one-pot and top-down synthetic route to produce brightly blue-emitting Au_8 NCs (quantum yield = 28.8%) protected by hyperbranched polyethyleneimine [103]. The fluorescence of Au_8 NCs was quenched upon the addition of hemoglobin with a detection limit of 5 nM. Moreover, a hemoglobin evaluation in blood samples with small relative standard deviations and satisfied recoveries was also achieved.

4.4   BIOLOGY AND BIOMEDICINE

The biological application of metal NCs is currently undergoing an impressive growth and new directions, including biolabeling [17], bioimaging [19], diagnosis [16], antimicrobial [104, 105], and cancer therapy [14, 17]. It is generally known that the low quantum yield and huge size of metal NPs hinder their applications in biology. Unlike metal NPs, metal NCs have exhibited better fluorescence in the broad region from visible to near-infrared with long lifetime, large Stokes shift, high compatibility, and superior photostablility [16]. Despite the high quantum yield (QY), small organic fluorescent dyes are easily bleached, exhibit small Stokes shifts and often have poor water solubility, all of which limit their usefulness in biolabeling and bioimaging [19]. Thus, metal NCs appear as better candidates in biological applications. More importantly, the smaller (or proper) sizes of the metal NCs lead to lower cytotoxicity and efficient renal clearance [14]. Another appealing feature of metal NCs is the capability of controlling properties, such as size and luminescence tunability, surface modification without compromising luminescence, better sensitivity to external environment. Besides, metal NCs are significantly less disruptive, for example, NC -labeled proteins can retain the biological functions of the proteins [106]. In a word, metal NCs exhibit distinct advantages in size, fluorescence, biocompatibility, and tunability for the application in biology than most of the other nanomaterials.

4.4.1 BIOLABELING AND BIOIMAGING

In 2005, Makarava et al. reported a type of water-soluble Ag NCs composed of Ag and thioflavin T and the imaging of amyloid fibrils was achieved using such Ag NCs [107]. Later, Yu et al. synthesized argyrophilic proteins scaffolded Ag NCs by intracellular photoactivation at ambient temperature [108]. The silver-stained cells can be monitored with picosecond-gated luminescence microscopy due to their short fluorescence lifetimes.

Lin et al. reported the synthesis of water-soluble fluorescent gold NCs capped with dihydrolipoic acid with a QY of 1–3% [109]. Then streptavidin-conjugated Au NCs via EDC coupling were used to label biotin inside human hepatoma cells (Fig. 4.12). Marjomäki et al. developed a precise protocol for site-specific conjugation of functionalized Au_{102} NCs to viral surfaces of enteroviruses echovirus 1 (EV1) and coxsackievirus B3 (CVB3) and the labeled viral particles were observed by TEM [110].

Liu et al. conducted vivo NIR fluorescence imaging of MCF-7 tumor-bearing mice using glutathione-coated luminescent Au NPs ($\sim$ 2.5 nm) and small dye molecules (IRDye 800CW) for comparison (Fig. 4.13) [111]. Notably, Au NCs showed a much longer tumor retention time and faster normal tissue clearance, indicating the well-known enhanced permeability and retention effect. Later the same group reported vivo fluorescence imaging of renal-clearable near-infrared-emitting Au NPs, indicating such low-cost fluorescence kidney functional imaging is promising for noninvasive monitoring or diagnosis of kidney dysfunction progression in preclinical research [112].

Yang et al. synthesized glutathione capped Ag_{14} NCs with red fluorescence [113]. The QY is revealed to be solvent-dependent with the assistance of cetyltrimethyl ammonium bromide. They applied Ag_{14} to label lung cancer cells (A549) for imaging successfully (Fig. 4.14) and also found the Ag_{14} NCs exhibit lower cytotoxicity compared with some other silver species (including silver salt, silver complex and large silver NPs). Later, the same group elucidated the structural origin of the strong fluorescence of Au_{24} NCs, which were further used for the imaging of living macrophages after surface modified by BSA [114]. Wang et al. developed a number-precise Ag-doping method and drastically enhanced the fluorescence of rod-shaped Au_{25} when

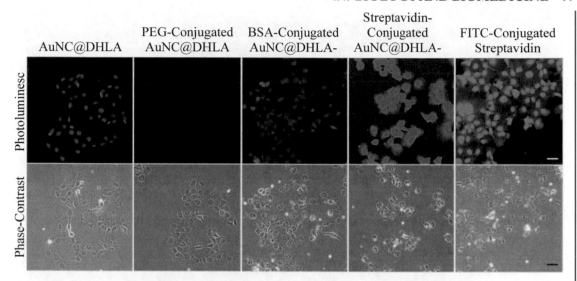

Figure 4.12: Labeling of endogenous biotin with fluorescent Au NCs conjugated to streptavidin in human hepatoma cells (HepG2). AuNC@DHLA:LP 585 em/488 ex, FITC:LP 530/488 ex. The scale bars indicate 50 μm. FITC-avidin was used as positive control. FITC: Fluorescein Isothiocyanate. Reproduced with permission from [109]. Copyright 2009 American Chemical Society.

the number of doped Ag atoms reached 13 [115]. By the introduction of hydroxyethylthiol, the $[Ag_xAu_{25-x}(PPh_3)_{10}(SC_2H_4OH)_5Cl_2]^{2+}$ ($x \leq 13$) NCs were applied to image in living cells (human cancer cell 7402).

4.4.2   DIAGNOSIS

Due to the fluorescent properties and resistance to photobleaching, a myriad of metal NCs have been used for the detection of small biomolecules, polypeptide, amino acids, and proteins, which are partly discussed in sensing application. In this subsection, we mainly introduce some diagnosis work using metal NCs with rational designing and demonstrate the application of metal NCs in vivo/vitro disease monitoring.

Hu et al. prepared a novel nanoprobe by the conjugation of folic acid to Au NCs protected by BSA [116]. The nanoprobe exhibited excellent stability, low cytotoxicity, high fluorescence and enzyme activity. The cancer can be quickly diagnosed after staining tumor tissue with this nanoprobes using microscopic imaging with bright field and fluorescent images simultaneously. The NCs-FA (where, FA = folic acid) nanoprobes strongly stained cancer cells while normal tissue samples showed negative staining. Further this nanoprobe can be used for monitoring

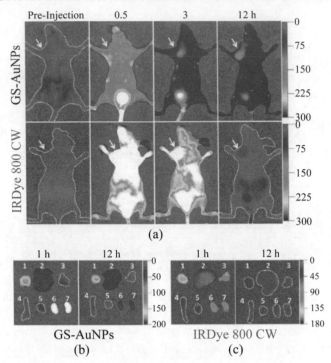

Figure 4.13: (a) NIR fluorescence images of MCF-7 tumor-bearing mice injected with Au NPs and IRDye 800CW. Ex vivo fluorescence images of organs and tumors removed 1 and 12 h p.i. from MCF-7 tumor-bearing mice injected with (b) Au NPs and (c) IRDye 800CW. Labels: 1, tumor; 2, liver; 3, lung; 4, spleen; 5, heart; 6, kidney (left); 7, kidney (right). Reproduced with permission from [111]. Copyright 2013 American Chemical Society.

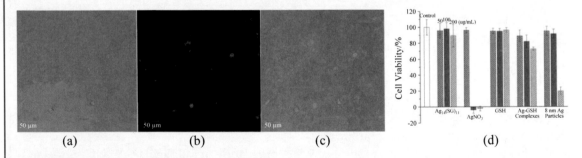

Figure 4.14: (a) The bright-field image, (b) TPF image (false color) upon excitation at 405/488 nm, and (c) the overlaid picture of lung cancer cells incubated with $Ag_{14}(SG)_{11}$ clusters for 24 h. (d) Comparison of the viability of A549 cells in Ag_{14}, $AgNO_3$, GSH, Ag-GSH complex, and 8 nm Ag NPs. Reproduced with permission from [113]. Copyright 2015 Royal Society of Chemistry.

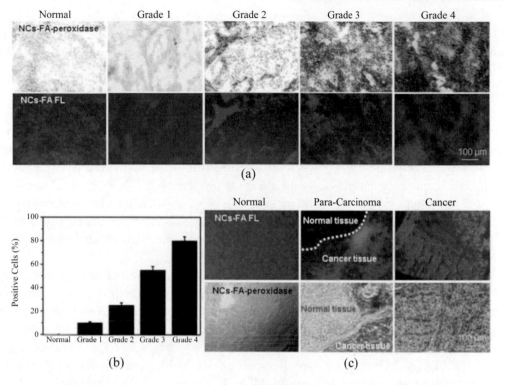

Figure 4.15: (a) The NCs-FA nanoprobes staining of a bladder cancer array. (b) Positive cells ratio in bladder cancer is associated with a more advanced stage of the disease. (c) The clinical liver cancer samples, para-carcinoma liver cancer tissue samples, and corresponding normal tissue samples were stained by the NCs-FA nanoprobes. Reproduced with permission from [116]. Copyright 2014 Ivyspring International Publisher.

the grade and growth pattern of bladder cancer samples for clinical application, as shown in Fig. 4.15.

Kurdekar et al. developed a gold NC immunoassay for the early and sensitive detection of HIV infection by conjugating streptavidin to AuSG NCs using EDC/sulfo-NHS chemistry [117]. The streptavidin-labeled Au NCs were applied for the detection of HIV-1 p24 antigen in clinical specimens with a sensitivity limit of 5 pg/mL based on the emitted fluorescent signals.

Loynachan et al. designed multifunctional protease nanosensors for the detection of colorectal cancer with a direct colorimetric urinary readout of the disease state [118]. The diagnosis can be achieved rapidly both in vitro and in vivo within one hour. The nanosensors were composed of Au NCs co-capped by glutathione and peptide via peptide linkages to a larger protein carrier (neutravidin, NAv). The AuNC-NAv complex will be disassembled in response to dysregulated protease activity at the site of disease and release free Au NCs. Then Au NCs can be filtered through the kidneys and excreted into the urine and finally detected by the oxidation of chromogenic peroxidase substrate. This work provides a versatile disease detection platform of catalytic nanomaterial diagnostics for a range of diseases.

4.4.3   ANTIMICROBIAL

Antibiotic resistance is considered as one of the greatest health threats worldwide due to the wide existence of bacteria [104]. Great efforts have demonstrated that metal NCs are promising antimicrobial agents owing to their high biocompatibility, ultrasmall size, higher surface to volume ratio, and superior physical and chemical properties [105]. Wang et al. reported a green and environment-friendly method for synthesis of water-soluble and fluorescent Ag NCs protected by carboxymethyl-β-cyclodextrin [119]. The Ag NCs exhibited the strongest antimicrobial ability in co-culture with a Gram-negative bacteria (E. coli) in comparison with the $AgNO_3$ solution and Ag NPs (Fig. 4.16).

Yuan et al. studied the antimicrobial properties of a strong luminescent $Ag_{16}(SG)_9$ NCs using P. aeruginosa [120]. The Ag_{16} NCs effectively inhibit the growth of P. aeruginosa even with the dosage of 8 μg. The Ag_{16} NCs even show better inhibition effect than that of chloramphenicol used at the same concentration. The intracellular generation of reactive oxygen species (ROS) has contributed to the superior antimicrobial activity (Fig. 4.17). Later, the same group found $Au_{25}(MHA)_{18}$ (MHA: 6-mercaptohexanoic acid) could kill both the Gram-positive and Gram-negative bacteria, which while can NOT be killed by Au NPs and Au(I)–MHA complexes [121]. Further, they investigated the antimicrobial ability of Au_{25} capped by different thiolate ligands. More negative –COOH groups of the ligands lead to more negatively charged surface for the preservation of more ROS, which can then kill more bacteria by interfering the normal metabolism of the bacteria.

Recently, Zheng et al. developed a synergistic and magnetically responsive antimicrobial agent by conjugating Au NCs onto Ho (holmium) modified GO nanosheets [122]. Since the assembled nanosheets can be vertically aligned under weak magnetic fields (< 0.5 T). They proposed that the formation of more sharp edges could effectively pierce the bacterial membrane and release Au NCs into bacteria, leading to the inhibition of the growth of both Gram-positive and Gram-negative bacteria.

4.4.4   CANCER THERAPY

Metal NCs as a type of emerging nanomaterials have received much attention in the application of cancer therapy due to their ultrasmall size with narrow size distribution and ease of conjugation [123]. Metal NCs can be effectively removed from the body via renal clearance. For example, Du et al. recently found glomerular barrier behaved as an atomically precise bandpass filter and only few-atom decreases in size could result in drastic reductions in renal clearance efficiency [124]. The same group further reported that glutathione efflux from hepatocytes could

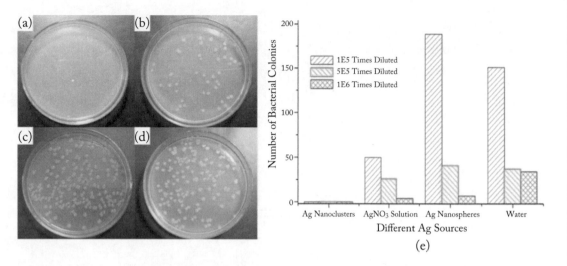

Figure 4.16: Antimicrobial activities of (a) Ag NCs, (b) AgNO$_3$, (c) Ag nanospheres, and (d) distilled water against E. coli. (e) The statistic results of bacterial colonies after cultured with different Ag sources. Reproduced with permission from [119]. Copyright 2013 Elsevier.

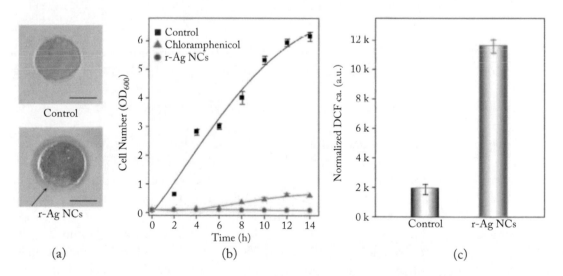

Figure 4.17: (a) Agar diffusion assay showing the presence of a clear zone surrounding the well where Ag$_{16}$ were introduced (scale bar of 1 cm). (b) Comparison in cell numbers between the control sample, the sample with chloramphenicol and the sample with Ag$_{16}$ after 14 h incubation. (c) Comparison of the ROS concentration of cells between the control sample and the Ag$_{16}$ sample. Reproduced with permission from [120]. Copyright 2013 Nature Publishing Group.

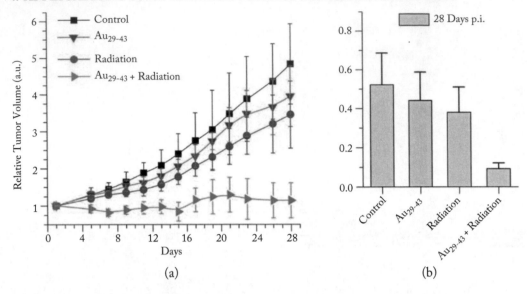

Figure 4.18: Time-course studies of tumor (a) volume and (b) weight of mice treated with $Au_{29-43}(SG)_{27-37}$ NCs at the concentration of 5.9 mg-Au/kg body. Reproduced with permission from [127]. Copyright 2015 Springer Nature.

transform the NCs surface chemistry and reduce their affinity to serum protein and significantly alters its blood retention, targeting and clearance, demonstrating metal NCs as a very promising candidate in nanomedicine and cancer therapy without the barrier of liver detoxification [125]. In the following, we will introduce some applications of metal NCs, especially those with atomic precision, in cancer therapy.

Zhang et al. reported the increased tumor uptake and targeting specificity for the ultra-small $Au_{10-12}(SG)_{10-12}$ NCs owing to their highly exposed biocompatible GSH shell [126]. As shown in Fig. 4.20, the ratios of the NCs concentration in tumor to that in kidney and liver were 1:0.172 and 1:0.0446, respectively. Therefore, the radiotherapy could be constrained within the tumors minimizing possible damages to normal tissues. The tumor volume after radiation in U14 tumor bearing mice treated with the NCs decreased $\sim$ 65% relative to tumors in the control group. Later they found $Au_{29-43}(SG)_{27-37}$ NCs performed well in the tumor uptake and targeting specificity like the case of $Au_{10-12}(SG)_{10-12}$ NCs [127]. A remarkable decrease (76%) of tumor volume was observed in mice treated with $Au_{29-43}(SG)_{27-37}$ NCs plus radiation, indicating that the glutathione capped Au NCs are attractive radiosensitizer materials for cancer radiotherapy (Fig. 4.18).

He et al. developed a new multifunctional cancer phototherapy platform by introducing captopril stabilized-Au NCs $Au_{25}(Capt)_{18}$ to the mesoporous silica shell of upconversion nanoparticles (UCNPs) [128]. Au_{25} NCs exhibit considerable photothermal effects and pho-

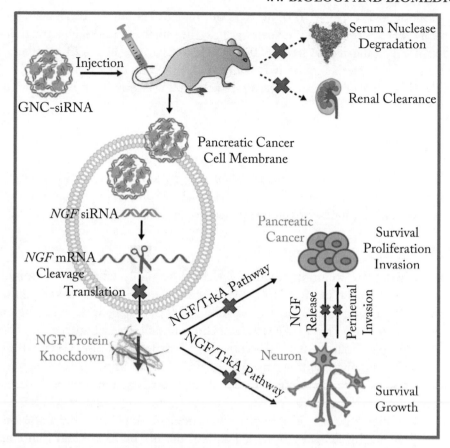

Figure 4.19: Delivery mechanism of siRNA-Gnanoclusters complex for NGF silencing and pancreatic cancer therapy. Reproduced with permission from [128]. Copyright 2017 The Authors.

tothermal imaging property. These nanomaterials also present the magnetic resonance and computer tomography imaging effects due to the existence of Gd^{3+} and Yb^{3+} ions in the UCNPs. The enhanced liver tumor (H22) inhibition ability was observed under 808-nm laser irradiation in vivo experiments. The body weight of all mice samples did not decrease with the prolonged time, indicating the nontoxicity of the material to the body.

Lei et al. reported a novel delivery platform for siRNA using siRNA-NCs (gold NCs) complex for pancreatic cancer treatment via nerve growth factors (NGF) depletion [129]. The complex potently downregulates the NGF expression in Panc-1 cells in vitro and in pancreatic tumors in vivo, leading to effective inhibition of tumor growth (Fig. 4.19).

Yang et al. deposited $Au_{25}(MHA)_{18}$ NCs on the PEG-modified black anatase TiO_{2-x} nanotubes by NHS/EDC coupling chemistry [130]. The enhanced photodynamic therapy was observed in tumor-bearing mouse upon 650 nm laser irradiation (Fig. 4.23). Katla et al. also reported the excellent phototherapy activity of $Au_{25}(SG)_{18}$ NCs [131]. In their case, 100% cell death of MDA-MB-231 breast cancer cells was achieved at a power of 10 W/cm^2 using an 808 nm laser source.

4.5   ENERGY

New materials are desirable for clean and affordable energy to reduce energy consumption and lessen toxicity on the environment [21]. Metal NCs have also been proved promising in electrochemical water splitting for hydrogen fuel generation, photovoltaic application, etc.

4.5.1   WATER SPLITTING

Electrochemical hydrogen evolution reaction (HER) is a highly efficient way for hydrogen generation using noble metal catalyst [132, 133]. Zhao et al. designed a new abundant catalyst by introducing Au NCs to molybdenum disulfide (MoS_2) nanosheets, resulting enhanced HER activity [134]. The $Au_{25}(PET)_{18}/MoS_2$ nanocomposite showed a higher current density of 59.3 mA cm^{-2} at the potential of -0.4 V than pure MoS_2 with a small onset potential of -0.20 V (Fig. 4.20). The HER enhancement by Au_{25} was attributed to the improved charge transfer and electronic interactions at both the cluster/MoS_2 and Au_{25} core/ligand interface. The same group also investigated the performance of Au NCs in oxygen evolution reaction (OER), which is a major bottleneck of high efficiency electrochemical water splitting [135]. The $CoSe_2$ supported Au_{25} exhibited enhanced OER activity with a current density of 10 mA cm^{-2} at small overpotential of ~ 0.43 V (Fig. 4.21). The OER activity increased with the cluster size increasing and the ligands lost. These studies may provide some guidelines for design of new catalysts for electrochemical water splitting.

4.5.2   PHOTOVOLTAICS

The pioneering work on utilization of metal NCs in the photovoltaic application was reported by Sakai et al. in 2010 [136]. They found Au_x-SG NCs adsorbed on TiO_2 electrodes could convert light to current under visible and/or near infrared light. In 2011, Zhang et al. developed an effective anode for polymer solar cells with Au NCs decorated multi-layer graphene [137]. Such anodes exhibited good performance after ultraviolet-ozone treatment with the power conversion efficiency up to 1.24%.

Au NCs have also been used as a new class of photosensitizers in mesoscopic TiO_2-based solar cells. By introducing Au_x-SG NCs in mesoscopic TiO_2 film, Chen et al. achieved relatively high power conversion efficiency (> 2%), with a stable photocurrent of 3.96 mA cm^{-2} under air mass 1.5 G (100 mW/cm^2) illumination (Table 4.8) [138]. In comparison with the cell

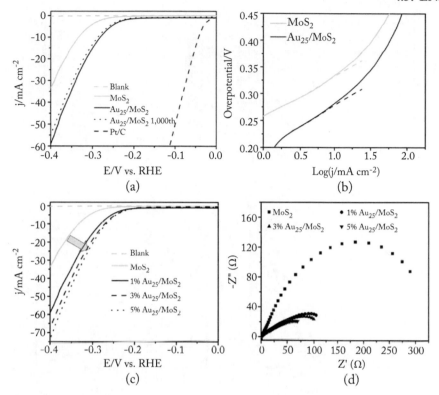

Figure 4.20: (a) HER polarization curves for Au_{25}/MoS_2, MoS_2, blank glassy carbon working electrode, Au_{25}/MoS_2 after the stability test, and commercial Pt/C catalysts. (b) Tafel plots of Au_{25}/MoS_2 and MoS_2 catalysts. (c) HER polarization curves and (d) impedance spectra for MoS_2 and Au_{25}/MoS_2 with different loading amounts of Au_{25}. Reproduced with permission from [134]. Copyright 2017 Wiley VCH.

performance of quantum dot analogs, metal cluster-sensitized solar cells show the viability for the next generation of solar cells. Later, the same group employed Au_x-SG NCs as a cosensitizer in dye-sensitized solar cells, and observed enhanced photon conversion efficiency and decreasing photovoltage penalty due to the dual roles of Au_x-SG NCs as a photosensitizer and voltage booster (Fig. 4.22) [139].

Abbas et al. explored the solar cell performance using $Au_x(SG)_y$ ($x = 10 - 25$, $y = 10 - 18$) NCs of various sizes with different number of core atoms fabricated on TiO_2 substrates [140]. The effect of Au NCs size on power conversion efficiency is not linear. Among them $Au_{18}(SR)_{14}$ achieved a best power conversion efficiency record of 3.8% due to the relatively good light absorption capability and low recombination rate (Fig. 4.23), which were suggested as the main factors in determining the solar cell performance. Recently, Yuan et al. reported

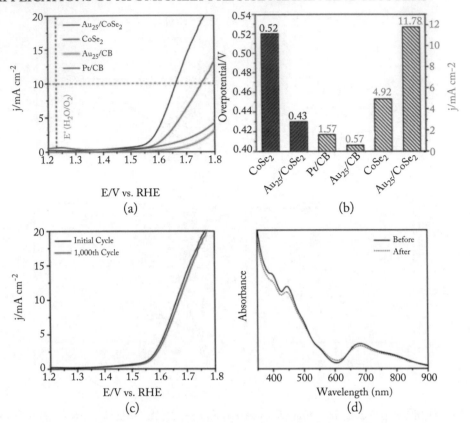

Figure 4.21: (a) OER polarization curves for $Au_{25}/CoSe_2$, $CoSe_2$, Pt/CB, and Au_{25}/CB. CB: carbon black. (b) Comparison of the overpotential required for achieving the current density of 10 mA cm^{-2}, and the current density at the overpotential of 0.45 V for $Au_{25}/CoSe_2$, $CoSe_2$, Pt/CB, and Au_{25}/CB catalysts. (c) OER polarization curves for $Au_{25}/CoSe_2$ before and after the stability test. (d) UV-vis spectra of Au_{25} before and after the stability test. Reproduced with permission from [135]. Copyright 2017 American Chemical Society.

the synthesis and structural determination of $[Cu_{53}(RCOO)_{10}(C \equiv CtBu)_{20}Cl_2H_{18}]^+$ (Cu_{53}) NCs [141]. They spin-coated Cu_{53} on organolead halide perovskite to form uniform NCs film and further converted it into high-quality CuI film by iodination. Such CuI film-fabricated perovskite solar cells showed a high power conversion efficiency of 14.3%, ignorable hysteresis, and high stability.

Table 4.8: Photovoltaic performance of solar cells fabricated by different sensitizers. Scheme adapted with permission from [138]. Copyright 2013 American Chemical Society. Data from [138].

Sensitizer	J_{sc} (mA cm^{-2})	V_{oc} (V)	ff	η(%)
Au$_x$-SG 1[a]	3.96	0.832	0.716	2.36
Au$_x$-SG 2[a]	3.50	0.825	0.701	2.03
Au$_x$-SG 3[a]	3.70	0.827	0.678	2.07
Au$_x$-SG 4[a]	3.82	0.809	0.681	2.11
Au$_x$-SG 5[a]	3.81	0.82	0.687	2.13
CdS	2.34	0.704	0.62	1.01
CdS/ZnS	7.52	0.537	0.579	2.34
None (TiO$_2$)	0.11	0.279	0.454	0.013

[a] Performance of five different Au$_x$-SG sensitized TiO$_2$ solar cells.

4.5.3 LIGHT-EMITTING DEVICE

Niesen et al. developed a novel thin film light-emitting device (LED) using metal NCs [142]. They incorporated Au or Ag NCs as emitters in a thin film LED structure. The electroluminescence (EL) measured from these devices is tunable by the choice of metal nanocluster (MNC) and also closely matched the photoluminescence (PL) of the MNC dispersions in solvent, demonstrating the MNC-LED as an additional option for future light-generating applications. Later Koh et al. reported the excellent performance in thin film light-emitting devices using Au$_x$-SG NCs as emitters, with luminance exceeding 40 cd m^{-2} and external quantum efficiency exceeding 0.1% with clearly visible orange emission [143].

Apart from Au NCs, Ag and Cu NCs have also been used for organic light emitting devices. Nishikitani synthesized a type of Ag NCs protected by poly(methacrylic acid) as color converters (yellow emitters) [144]. When cooperating with polymer light-emitting electrochemi-

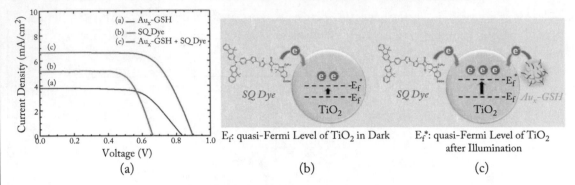

Figure 4.22: (a) J–V characteristics of solar cell recorded under AM 1.5 G illumination (100 mW/cm^2) using Au$_x$-SG, SQ dye, and Au$_x$-SG + SQ dye-sensitized TiO$_2$ as photoanode. (b) and (c) schematic illustration of the increase in quasi-Fermi level of TiO$_2$ after illumination. Reproduced with permission from [139]. Copyright 2015 American Chemical Society.

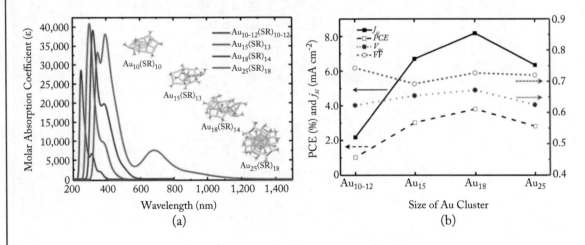

Figure 4.23: (a) Simulated UV-vis spectra of Au NCs obtained from TD-DFT calculations. (b) Summary of the solar cell parameters of the Au NC-sensitized solar cells. Reproduced with permission from [140]. Copyright 2016 American Chemical Society.

cal cells fabricated with a blue fluorescent p-conjugated polymer (blue emitters), pure white light emission with Commission Internationale de l'Eclairage (CIE) coordinates of ($x = 0.32$, $y = 0.33$) and a color rendering index (CRI) of 93.6 were successfully achieved (Fig. 4.24). Wang et al. developed a class of white light-emitting devices (WLED) solely based on Cu NCs to meet the problems of the cost and shortage in supply of commonly utilized rare-earth elements for WLEDs [145]. Orange-emitting Cu NCs capped by SG and blue-emitting Cu NCs

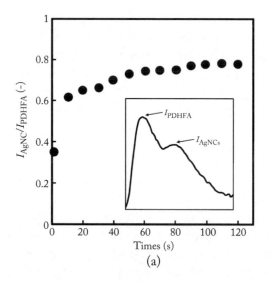
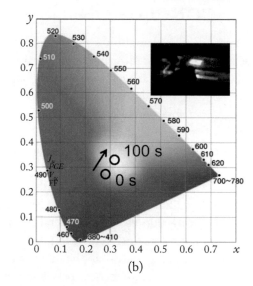

Figure 4.24: (a) Temporal evolution of the ratio of I_{AgNCs} to $I_{polymer}$, where $I_{polymer}$ is the peak EL intensity of the polymer light-emitting electrochemical cells without the Ag NC color converter and I_{AgNCs} is the peak PL intensity of the Ag NC color converter. (b) Change in CIE coordinates of the emission from the polymer light-emitting electrochemical cells with the Ag NC color converter. The devices were operated at a constant voltage of 13 V. Reproduced with permission from [144]. Copyright 2014 American Institute of Physics.

protected by polyvinylpyrrolidone were integrated on a commercial GaN LED chip achieved a white light with CIE and CRI of (0.36, 0.31) and 92, respectively.

4.5.4   OTHER ENERGY CONVERSION

Apparently, employing metal NCs as a photosensitizer or a light harvester in energy conversion applications is one of the rapidly emerging research areas. Considering effective and biological utilization of solar power, Zhang et al. recently reported the construction of a newly photosynthetic biohybrid systems (PBSs) by introducing Au NCs as the biocompatible intracellular photosensitizer for non-photosynthetic bacteria [146]. Such a platform can efficiently harvest sunlight and transfer photogenerated electrons to cellular metabolism for CO_2 fixation. Incorporation of $Au_{22}(SG)_{18}$ into non-photosynthetic bacteria (Moorella thermoacetica) enables photosynthesis of acetic acid from CO_2. Au_{22} NCs also inhibited the formation of reactive oxygen species during the photo-irradiation, maintaining the high bacterium viability and continuous CO_2 fixation over six days.

4.6   CONCLUSION

In the past decade, metal NCs have emerged as a new class of nanomaterials with many unique properties, such as the ultrasmall size, precise composition and structure, attractive optical properties, structure diversity and tunability, and biocompatibility. The catalytic performance of metal NCs is quite unique, as such materials exhibit not only high activity but also excellent selectivity in a wide range of reactions. More importantly, precisely tuning the composition and structure of metal NCs contributes largely to the unraveling of catalytic mechanisms.

For future work, deeper understanding of the catalytic mechanism is still desirable. With the aid of the analytical tools (such as X-ray crystallography, mass spectrometry, optical spectroscopy, etc.) and the synthetic methods, as well as theoretical calculations, mechanistic studies are expected to gain more progress using well-defined metal NC catalysts.

The strong fluorescence of metal NCs will contribute more to the sensor development such as strain sensors and multi-modular ones. The development of multifunctional sensors is quite attractive for practical applications, for example, the integration of the responses of the magnetism, photoluminescence, photothermal effect, and the light absorption using versatile NC materials.

Metal NCs are still in their infancy in the biology and biomedicine sectors, however, attractive performance has been reported in labeling, diagnosis, antimicrobial and therapy, among others. For intracellular targeted nanodelivery and nanomedicine system, the surface functionalization of metal NCs remains crucial for biological and medicinal applications in enhancing the stability, interaction, and transport, and targeting specificity. In addition to luminescent NCs, the usage of non-fluorescent metal NCs is still a challenge in terms of adequate traceability inside live cells or organism.

In various energy applications, metal NCs with discrete energy levels and unique photophysical properties have also shown great promise. These materials can serve as a new light harvester. While noble metal NCs (such as Au, Pt, Pd) for practical energy conversion may be of concern because of their high cost and insufficient abundances, the fundamental mechanisms learned will lead to the design of nonprecious metal or alloy NCs to resolve the issues. The conjugation of precious metal NCs to other cheaper materials with enhanced efficiency would be viable for the practical usage in energy.

In summary, atomically precise metal NCs are expected to have a great leap in future research, not only in fundamental science research but also in practical applications.

4.7   REFERENCES

[1] Templeton, A. C., Wuelfing, W. P., and Murray, R. W. Monolayer-protected cluster molecules. *Acc. Chem. Res.*, 33:27–36, 2000. DOI: 10.1021/ar9602664. 79

[2] Jin, R., Zeng, C., Zhou, M., and Chen, Y. Atomically precise colloidal metal nanoclusters and nanoparticles: Fundamentals and opportunities. *Chem. Rev.*, 116:10346–10413,

2016. DOI: 10.1021/acs.chemrev.5b00703. 79, 80

[3] Chakraborty, I. and Pradeep, T. Atomically precise clusters of noble metals: Emerging link between atoms and nanoparticles. *Chem. Rev.*, 117:8208–8271, 2017. DOI: 10.1021/acs.chemrev.6b00769. 79

[4] Yao, Q., Yuan, X., Chen, T., Leong, D. T., and Xie, J. Metal nanoclusters: Engineering functional metal materials at the atomic level. *Adv. Mater.*, 30:1870358, 2018. DOI: 10.1002/adma.201870358. 79

[5] Kang, X., Chong, H., and Zhu, M. $Au_{25}(SR)_{18}$: The captain of the great nanocluster ship. *Nanoscale*, 10:10758–10834, 2018. DOI: 10.1039/c8nr02973c. 79, 80

[6] Li, G. and Jin, R. Atomically precise gold nanoclusters as new model catalysts. *Acc. Chem. Res.*, 46:1749–1758, 2013. DOI: 10.1021/ar300213z. 79, 80

[7] Higaki, T., Li, Y., Zhao, S., Li, Q., Li, S., Du, X.-S., Yang, S., Chai, J., and Jin, R. Atomically tailored gold nanoclusters for catalytic application. *Angew. Chem. Int. Ed.*, 58:8291–8302, 2019. DOI: 10.1002/anie.201814156. 79, 80

[8] Fang, J., Zhang, B., Yao, Q., Yang, Y., Xie, J., and Yan, N. Recent advances in the synthesis and catalytic applications of ligand-protected, atomically precise metal nanoclusters. *Coord. Chem. Rev.*, 322:1–29, 2016. DOI: 10.1016/j.ccr.2016.05.003. 79, 80

[9] Diez, I. and Ras, R. H. A. Fluorescent silver nanoclusters. *Nanoscale*, 3:1963–1970, 2011. DOI: 10.1039/c1nr00006c. 79

[10] Kang, X. and Zhu, M. Tailoring the photoluminescence of atomically precise nanoclusters. *Chem. Soc. Rev.*, 48:2422–2457, 2019. DOI: 10.1039/c8cs00800k. 79

[11] Yu, H., Rao, B., Jiang, W., Yang, S., and Zhu, M. The photoluminescent metal nanoclusters with atomic precision. *Coord. Chem. Rev.*, 378:595–617, 2019. DOI: 10.1016/j.ccr.2017.12.005. 79

[12] Mathew, A. and Pradeep, T. Noble metal clusters: Applications in energy, environment, and biology. *Part. Syst. Charact.*, 31:1017–1053, 2014. DOI: 10.1002/ppsc.201400033. 79

[13] Yu, M. and Zheng, J. Clearance pathways and tumor targeting of imaging nanoparticles. *ACS Nano*, 9:6655–6674, 2015. DOI: 10.1021/acsnano.5b01320. 79

[14] Zuber, G., Weiss, E., and Chiper, M. Biocompatible gold nanoclusters: Synthetic strategies and biomedical prospects. *Nanotechnology*, 30:352001, 2019. DOI: 10.1088/1361-6528/ab2088. 79, 97, 98

[15] Goswami, N., Zheng, K., and Xie, J. Bio-NCs—the marriage of ultrasmall metal nanoclusters with biomolecules. *Nanoscale*, 6:13328–13347, 2014. DOI: 10.1039/c4nr04561k. 79

[16] Tao, Y., Li, M., Ren, J., and Qu, X. Metal nanoclusters: Novel probes for diagnostic and therapeutic applications. *Chem. Soc. Rev.*, 44:8636–8663, 2015. DOI: 10.1039/c5cs00607d. 79, 97

[17] Su, Y., Xue, T., Liu, Y., Qi, J., Jin, R., and Lin, Z. Luminescent metal nanoclusters for biomedical applications. *Nano Res.*, 12:1251–1265, 2019. DOI: 10.1007/s12274-019-2314-y. 79, 97

[18] Kaur, N., Aditya, R. N., Singh, A., and Kuo, T.-R. Biomedical applications for gold nanoclusters: Recent developments and future perspectives. *Nanoscale Res. Lett.*, 13:302, 2018. DOI: 10.1186/s11671-018-2725-9. 79

[19] Zhang, Y., Zhang, C., Xu, C., Wang, X., Liu, C., Waterhouse, G. I. N., Wang, Y., and Yin, H. Ultrasmall Au nanoclusters for biomedical and biosensing applications: A mini-review. *Talanta*, 200:432–442, 2019. DOI: 10.1016/j.talanta.2019.03.068. 79, 97, 98

[20] Luo, Z., Zheng, K., and Xie, J. Engineering ultrasmall water-soluble gold and silver nanoclusters for biomedical applications. *Chem. Commun.*, 50:5143–5155, 2014. DOI: 10.1039/c3cc47512c. 79, 80

[21] Munir, A., Joya, K. S., Ul Haq, T., Babar, N.-U.-A., Hussain, S. Z., Qurashi, A., Ul-lah, N., and Hussain, I. Metal nanoclusters: New paradigm in catalysis for water splitting, solar and chemical energy conversion. *ChemSusChem*, 12:1517–1548, 2019. DOI: 10.1002/cssc.201802069. 79, 106

[22] Zhou, M., Zeng, C., Chen, Y., Zhao, S., Sfeir, M. Y., Zhu, M., and Jin, R. Evolution from the plasmon to exciton state in ligand-protected atomically precise gold nanoparticles. *Nature Communications*, 7:13240, 2016. DOI: 10.1038/ncomms13240. 79, 81

[23] Hossain, S., Niihori, Y., Nair, L. V., Kumar, B., Kurashige, W., and Negishi, Y. Alloy clusters: Precise synthesis and mixing effects. *Acc. Chem. Res.*, 51:3114–3124, 2018. DOI: 10.1021/acs.accounts.8b00453. 79

[24] Gan, Z., Xia, N., and Wu, Z. Discovery, mechanism, and application of antigalvanic reaction. *Acc. Chem. Res.*, 51:2774–2783, 2018. DOI: 10.1021/acs.accounts.8b00374. 79

[25] Yan, J., Teo, B. K., and Zheng, N. Surface chemistry of atomically precise coinage—metal nanoclusters: From structural control to surface reactivity and catalysis. *Acc. Chem. Res.*, 51:3084–3093, 2018. DOI: 10.1021/acs.accounts.8b00371. 79

[26] Wu, Z. and Jin, R. On the ligand's role in the fluorescence of gold nanoclusters. *Nano Lett.*, 10:2568–2573, 2010. DOI: 10.1021/nl101225f. 79

[27] Higaki, T., Li, Q., Zhou, M., Zhao, S., Li, Y., Li, S., and Jin, R. Toward the tailoring chemistry of metal nanoclusters for enhancing functionalities. *Acc. Chem. Res.*, 51:2764–2773, 2018. DOI: 10.1021/acs.accounts.8b00383. 79

[28] Zhao, J. and Jin, R. Heterogeneous catalysis by gold and gold-based bimetal nanoclusters. *Nano Today*, 18:86–102, 2018. DOI: 10.1016/j.nantod.2017.12.009. 80, 87

[29] Masatake, H., Tetsuhiko, K., Hiroshi, S., and Nobumasa, Y. Novel gold catalysts for the oxidation of carbon monoxide at a temperature far below 0°C. *Chem. Lett.*, 16:405–408, 1987. DOI: 10.1246/cl.1987.405. 80

[30] Gao, F., Wood, T. E., and Goodman, D. W. The effects of water on CO oxidation over TiO_2 supported Au catalysts. *Catal. Lett.*, 134:9–12, 2010. DOI: 10.1007/s10562-009-0224-4. 80

[31] Nie, X., Qian, H., Ge, Q., Xu, H., and Jin, R. CO oxidation catalyzed by oxide-supported $Au_{25}(SR)_{18}$ nanoclusters and identification of perimeter sites as active centers. *ACS Nano*, 6:6014–6022, 2012. DOI: 10.1021/nn301019f. 80, 81

[32] Haruta, M. Spiers memorial lecture role of perimeter interfaces in catalysis by gold nanoparticles. *Faraday Discuss.*, 152:11–32, 2011. DOI: 10.1039/c1fd00107h. 80

[33] Laha, S. C. and Kumar, R. Selective epoxidation of styrene to styrene oxide over TS-1 using urea—hydrogen peroxide as oxidizing agent. *J. Catal.*, 204:64–70, 2001. DOI: 10.1006/jcat.2001.3352. 81

[34] Turner, M., Golovko, V. B., Vaughan, O. P. H., Abdulkin, P., Berenguer-Murcia, A., Tikhov, M. S., Johnson, B. F. G., and Lambert, R. M. Selective oxidation with dioxygen by gold nanoparticle catalysts derived from 55-atom clusters. *Nature*, 454:981–983, 2008. DOI: 10.1038/nature07194. 82

[35] Zhu, Y., Qian, H., Zhu, M., and Jin, R. Thiolate-protected Au_n nanoclusters as catalysts for selective oxidation and hydrogenation processes. *Adv. Mater.*, 22:1915–1920, 2010. DOI: 10.1002/adma.200903934. 82

[36] Liu, Y., Tsunoyama, H., Akita, T., and Tsukuda, T. Efficient and selective epoxidation of styrene with TBHP catalyzed by Au_{25} clusters on hydroxyapatite. *Chem. Commun.*, 46:550–552, 2010. DOI: 10.1039/b921082b. 82, 83

[37] Wang, S., Jin, S., Yang, S., Chen, S., Song, Y., Zhang, J., and Zhu, M. Total structure determination of surface doping $[Ag_{46}Au_{24}(SR)_{32}](BPh_4)_2$ nanocluster and its structure-related catalytic property. *Sci. Adv.*, 1:e1500441, 2015. DOI: 10.1126/sciadv.1500441. 82

[38] Tsunoyama, H., Ichikuni, N., Sakurai, H., and Tsukuda, T. Effect of electronic structures of Au clusters stabilized by poly(N-vinyl-2-pyrrolidone) on aerobic oxidation catalysis. *J. Am. Chem. Soc.*, 131:7086–7093, 2009. DOI: 10.1021/ja810045y. 82

[39] Yoskamtorn, T., Yamazoe, S., Takahata, R., Nishigaki, J.-i., Thivasasith, A., Limtrakul, J., and Tsukuda, T. Thiolate-mediated selectivity control in aerobic alcohol oxidation by porous carbon-supported Au_{25} clusters. *ACS Catal.*, 4:3696–3700, 2014. DOI: 10.1021/cs501010x. 83

[40] Wang, L., Chai, X., Cheng, X., and Zhu, Y. Structure-specific catalytic oxidation with O2 by isomers in $Au_{28}(SR)_{20}$ nanoclusters. *ChemistrySelect*, 3:6165–6169, 2018. DOI: 10.1002/slct.201800914. 84

[41] Liu, Y., Tsunoyama, H., Akita, T., Xie, S., and Tsukuda, T. Aerobic oxidation of cyclohexane catalyzed by size-controlled Au clusters on hydroxyapatite: Size effect in the sub-2 nm regime. *ACS Catal.*, 1:2–6, 2011. DOI: 10.1021/cs100043j. 84, 85

[42] Tian, S., Li, Y. Z., Li, M. B., Yuan, J., Yang, J., Wu, Z., and Jin, R. Structural isomserism in gold nanoparticles revealed by X-ray crystallography. *Nat. Commun.*, 6:10012, 2015. DOI: 10.1038/ncomms9667. 84

[43] Li, M.-B., Tian, S.-K., Wu, Z., and Jin, R. Cu^{2+} induced formation of $Au_{44}(SC_2H_4Ph)_{32}$ and its high catalytic activity for the reduction of 4-nitrophenol at low temperature. *Chem. Commun.*, 51:4433–4436, 2015. DOI: 10.1039/c4cc08830a. 84, 85

[44] Li, G. and Jin, R. Gold nanocluster-catalyzed semihydrogenation: A unique activation pathway for terminal alkynes. *J. Am. Chem. Soc.*, 136:11347–11354, 2014. DOI: 10.1021/ja503724j. 84, 86

[45] Wan, X.-K., Wang, J.-Q., Nan, Z.-A., and Wang, Q.-M. Ligand effects in catalysis by atomically precise gold nanoclusters. *Sci. Adv.*, 3:e1701823, 2017. DOI: 10.1126/sciadv.1701823. 84

[46] Chen, H., Li, Z., Qin, Z., Kim, H. J., Abroshan, H., and Li, G. Silica-encapsulated gold nanoclusters for efficient acetylene hydrogenation to ethylene. *ACS Appl. Nano Mater.*, 2:2999–3006, 2019. DOI: 10.1021/acsanm.9b00384. 86

[47] Zhu, Y., Qian, H., Drake, B. A., and Jin, R. Atomically precise $Au_{25}(SR)_{18}$ nanoparticles as catalysts for the selective hydrogenation of α, β-unsaturated ketones and aldehydes. *Angew. Chem. Int. Ed.*, 49:1295–1298, 2010. DOI: 10.1002/anie.200906249. 86, 87

[48] Zhu, Y., Wu, Z., Gayathri, C., Qian, H., Gil, R. R., and Jin, R. Exploring stereoselectivity of Au_{25} nanoparticle catalyst for hydrogenation of cyclic ketone. *J. Catal.*, 271:155–160, 2010. DOI: 10.1016/j.jcat.2010.02.027. 86

[49] Li, M.-B., Tian, S.-K., and Wu, Z. Catalyzed formation of α, β-unsaturated ketones or aldehydes from propargylic acetates by a recoverable and recyclable nanocluster catalyst. *Nanoscale*, 6:5714–5717, 2014. DOI: 10.1039/c4nr00658e. 87, 88

[50] Li, G. and Jin, R., Catalysis by gold nanoparticles: Carbon-carbon coupling reactions. *Nanotechnol. Rev.*, 2:529, 2013. DOI: 10.1515/ntrev-2013-0020. 87

[51] Shi, Q., Qin, Z., Xu, H., and Li, G. Heterogeneous cross-coupling over gold nanoclusters. *Nanomaterials*, 9:838, 2019. DOI: 10.3390/nano9060838. 87

[52] Li, G., Liu, C., Lei, Y., and Jin, R. Au_{25} nanocluster-catalyzed Ullmann-type homocoupling reaction of aryl iodides. *Chem. Commun.*, 48:12005–12007, 2012. DOI: 10.1039/c2cc34765b. 87

[53] Li, G., Abroshan, H., Liu, C., Zhuo, S., Li, Z., Xie, Y., Kim, H. J., Rosi, N. L., and Jin, R. Tailoring the electronic and catalytic properties of Au_{25} nanoclusters via ligand engineering. *ACS Nano*, 10:7998–8005, 2016. DOI: 10.1021/acsnano.6b03964. 87

[54] Abroshan, H., Li, G., Lin, J., Kim, H. J., and Jin, R. Molecular mechanism for the activation of $Au_{25}(SCH_2CH_2Ph)_{18}$ nanoclusters by imidazolium-based ionic liquids for catalysis. *J. Catal.*, 337:72–79, 2016. DOI: 10.1016/j.jcat.2016.01.011. 87

[55] Li, G., Jiang, D.-E., Liu, C., Yu, C., and Jin, R. Oxide-supported atomically precise gold nanocluster for catalyzing sonogashira cross-coupling. *J. Catal.*, 306:177–183, 2013. DOI: 10.1016/j.jcat.2013.06.017. 87

[56] Li, Q., Das, A., Wang, S., Chen, Y., and Jin, R. Highly efficient three-component coupling reaction catalysed by atomically precise ligand-protected $Au_{38}(SC_2H_4Ph)_{24}$ nanoclusters. *Chem. Commun.*, 52:14298–14301, 2016. DOI: 10.1039/c6cc07825g. 89

[57] Adachi, Y., Kawasaki, H., Nagata, T., and Obora, Y. Thiolate-protected gold nanoclusters $Au_{25}(phenylethanethiol)_{18}$: An efficient catalyst for the synthesis of propargylamines from aldehydes, amines, and alkynes. *Chem. Lett.*, 45:1457–1459, 2016. DOI: 10.1246/cl.160813. 89

[58] Li, M.-B., Tian, S.-k., and Wu, Z. Improving the catalytic activity of Au_{25} nanocluster by peeling and doping. *Chin. J. Chem.*, 35:567–571, 2017. DOI: 10.1002/cjoc.201600526. 89

[59] Yao, C., Chen, J., Li, M.-B., Liu, L., Yang, J., and Wu, Z. Adding two active silver atoms on Au_{25} nanoparticle. *Nano Lett.*, 15:1281–1287, 2015. DOI: 10.1021/nl504477t. 89

[60] Kogo, A., Sakai, N., and Tatsuma, T. Photocatalysis of Au_{25}-modified TiO_2 under visible and near infrared light. *Electrochem. Commun.*, 12:996–999, 2010. DOI: 10.1016/j.elecom.2010.05.021. 90

[61] Yu, C., Li, G., Kumar, S., Kawasaki, H., and Jin, R. Stable $Au_{25}(SR)_{18}/TiO_2$ composite nanostructure with enhanced visible light photocatalytic activity. *J. Phys. Chem. Lett.*, 4:2847–2852, 2013. DOI: 10.1021/jz401447w. 90

[62] Li, Z., Liu, C., Abroshan, H., Kauffman, D. R., and Li, G. $Au_{38}S_2(SAdm)_{20}$ photocatalyst for one-step selective aerobic oxidations. *ACS Catal.*, 7:3368–3374, 2017. DOI: 10.1021/acscatal.7b00239. 90

[63] Chen, H., Liu, C., Wang, M., Zhang, C., Luo, N., Wang, Y., Abroshan, H., Li, G., and Wang, F. Visible light gold nanocluster photocatalyst: Selective aerobic oxidation of amines to imines. *ACS Catal.*, 7:3632–3638, 2017. DOI: 10.1021/acscatal.6b03509. 90, 91

[64] Liu, C., Ren, X., Lin, F., Fu, X., Lin, X., Li, T., Sun, K., and Huang, J. Structure of the $Au_{23-x}Ag_x(S\text{-}Adm)_{15}$ nanocluster and its application for photocatalytic degradation of organic pollutants. *Angew. Chem. Int. Ed.*, 58:11335–11339, 2019. DOI: 10.1002/anie.201904612. 91

[65] Kwak, K. and Lee, D. Electrochemistry of atomically precise metal nanoclusters. *Acc. Chem. Res.*, 52:12–22, 2019. DOI: 10.1021/acs.accounts.8b00379. 91

[66] Chen, W. and Chen, S. Oxygen electroreduction catalyzed by gold nanoclusters: Strong core size effects. *Angew. Chem. Int. Ed.*, 48:4386–4389, 2009. DOI: 10.1002/anie.200901185. 91

[67] Kauffman, D. R., Alfonso, D., Matranga, C., Qian, H., and Jin, R. Experimental and computational investigation of Au_{25} clusters and CO_2: A unique interaction and enhanced electrocatalytic activity. *J. Am. Chem. Soc.*, 134:10237–10243, 2012. DOI: 10.1021/ja303259q. 91, 92

[68] Zhao, S., Austin, N., Li, M., Song, Y., House, S. D., Bernhard, S., Yang, J. C., Mpourmpakis, G., and Jin, R. Influence of atomic-level morphology on catalysis: The case of sphere and rod-like gold nanoclusters for CO_2 electroreduction. *ACS Catal.*, 8:4996–5001, 2018. DOI: 10.1021/acscatal.8b00365. 92

[69] Lu, Y., Zhang, C., Li, X., Frojd, A. R., Xing, W., Clayborne, A. Z., and Chen, W. Significantly enhanced electrocatalytic activity of Au_{25} clusters by single platinum atom doping. *Nano Energy*, 50:316–322, 2018. DOI: 10.1016/j.nanoen.2018.05.052. 92

[70] Zhuang, S., Chen, D., Liao, L., Zhao, Y., Xia, N., Zhang, W., Wang, C., Yang, J., and Wu, Z. Hard-sphere random close-packed $Au_{47}Cd_2(TBBT)_{31}$ nanoclusters with a faradaic efficiency of up to 96% for electrocatalytic CO_2 reduction to CO. *Angew. Chem. Int. Ed.*, 59:3073–3077, 2020. DOI: 10.1002/anie.201912845. 93

[71] Saha, K., Agasti, S. S., Kim, C., Li, X., and Rotello, V. M. Gold nanoparticles in chemical and biological sensing. *Chem. Rev.*, 112:2739–2779, 2012. DOI: 10.1021/cr2001178. 93

[72] Guo, C. and Irudayaraj, J. Fluorescent Ag clusters via a protein-directed approach as a Hg(II) ion sensor. *Anal. Chem.*, 83:2883–2889, 2011. DOI: 10.1021/ac1032403. 93

[73] Huang, C.-C., Yang, Z., Lee, K.-H., and Chang, H.-T. Synthesis of highly fluorescent gold nanoparticles for sensing mercury(II). *Angew. Chem. Int. Ed.*, 46:6824–6828, 2007. DOI: 10.1002/anie.200700803. 93

[74] Shang, L. and Dong, S. Silver nanocluster-based fluorescent sensors for sensitive detection of Cu(ii). *J. Mater. Chem.*, 18:4636–4640, 2008. DOI: 10.1039/b810409c. 93

[75] Roy, S., Palui, G., and Banerjee, A. The as-prepared gold cluster-based fluorescent sensor for the selective detection of As III ions in aqueous solution. *Nanoscale*, 4:2734–2740, 2012. DOI: 10.1039/c2nr11786j. 93

[76] Zhang, J., Yuan, Y., Liang, G., Arshad, M. N., Albar, H. A., Sobahi, T. R., and Yu, S.-H. A microwave-facilitated rapid synthesis of gold nanoclusters with tunable optical properties for sensing ions and fluorescent ink. *Chem. Commun.*, 51:10539–10542, 2015. DOI: 10.1039/c5cc03086b. 93

[77] Habeeb Muhammed, M. A., Verma, P. K., Pal, S. K., Retnakumari, A., Koyakutty, M., Nair, S., and Pradeep, T. Luminescent quantum clusters of gold in bulk by albumin-induced core etching of nanoparticles: Metal ion sensing, metal-enhanced luminescence, and biolabeling. *Chem. Eur. J.*, 16:10103–10112, 2010. DOI: 10.1002/chem.201000841. 93

[78] Zheng, B., Zheng, J., Yu, T., Sang, A., Du, J., Guo, Y., Xiao, D., and Choi, M. M. F. Fast microwave-assisted synthesis of AuAg bimetallic nanoclusters with strong yellow emission and their response to mercury(II) ions. *Sens. Actuators B Chem.*, 221:386–392, 2015. DOI: 10.1016/j.snb.2015.06.089. 93

[79] Xie, J., Zheng, Y., and Ying, J. Y. Highly selective and ultrasensitive detection of Hg^{2+} based on fluorescence quenching of Au nanoclusters by Hg^{2+}–Au^+ interactions. *Chem. Commun.*, 46:961–963, 2010. DOI: 10.1039/b920748a. 93

[80] Chen, W.-Y., Lan, G.-Y., and Chang, H.-T. Use of fluorescent DNA-templated gold/silver nanoclusters for the detection of sulfide ions. *Anal. Chem.*, 83:9450–9455, 2011. DOI: 10.1021/ac202162u. 93

[81] Zhang, J., Chen, C., Xu, X., Wang, X., and Yang, X. Use of fluorescent gold nanoclusters for the construction of a NAND logic gate for nitrite. *Chem. Commun.*, 49:2691–2693, 2013. DOI: 10.1039/c3cc38298b. 93

[82] Chen, W.-Y., Huang, C.-C., Chen, L.-Y., and Chang, H.-T. Self-assembly of hybridized ligands on gold nanodots: tunable photoluminescence and sensing of nitrite. *Nanoscale*, 6:11078–11083, 2014. DOI: 10.1039/c4nr02817a. 93

[83] Li, Z., Guo, S., and Lu, C. A highly selective fluorescent probe for sulfide ions based on aggregation of Cu nanocluster induced emission enhancement. *Analyst*, 140:2719–2725, 2015. DOI: 10.1039/c5an00017c. 93

[84] Wen, F., Dong, Y., Feng, L., Wang, S., Zhang, S., and Zhang, X. Horseradish peroxidase functionalized fluorescent gold nanoclusters for hydrogen peroxide sensing. *Anal. Chem.*, 83:1193–1196, 2011. DOI: 10.1021/ac1031447. 93

[85] Yuan, X., Tay, Y., Dou, X., Luo, Z., Leong, D. T., and Xie, J. Glutathione-protected silver nanoclusters as cysteine-selective fluorometric and colorimetric probe. *Anal. Chem.*, 85:1913–1919, 2013. DOI: 10.1021/ac3033678. 93

[86] Liu, Y., Ai, K., Cheng, X., Huo, L., and Lu, L. Gold-nanocluster-based fluorescent sensors for highly sensitive and selective detection of cyanide in water. *Adv. Funct. Mater.*, 20:951–956, 2010. DOI: 10.1002/adfm.200902062. 93

[87] Song, W., Wang, Y., Liang, R.-P., Zhang, L., and Qiu, J.-D. Label-free fluorescence assay for protein kinase based on peptide biomineralized gold nanoclusters as signal sensing probe. *Biosens. Bioelectron.*, 64:234–240, 2015. DOI: 10.1016/j.bios.2014.08.082. 93

[88] Sharma, J., Yeh, H.-C., Yoo, H., Werner, J. H., and Martinez, J. S. Silver nanocluster aptamers: In situ generation of intrinsically fluorescent recognition ligands for protein detection. *Chem. Commun.*, 47:2294–2296, 2011. DOI: 10.1039/C0CC03711G. 93

[89] Zhou, Y., Zhou, T., Zhang, M., and Shi, G. A DNA–scaffolded silver nanocluster/Cu^{2+} ensemble as a turn-on fluorescent probe for histidine. *Analyst*, 139:3122–3126, 2014. DOI: 10.1039/c4an00487f. 93

[90] Yu, Q., Gao, P., Zhang, K. Y., Tong, X., Yang, H., Liu, S., Du, J., Zhao, Q., and Huang, W. Luminescent gold nanocluster-based sensing platform for accurate H_2S detection *in vitro* and *in vivo* with improved anti-interference. *Light Sci. Appl.*, 6:e17107, 2017. DOI: 10.1038/lsa.2017.107. 93

[91] Wang, Y., Wang, Y., Zhou, F., Kim, P., and Xia, Y. Protein-protected Au clusters as a new class of nanoscale biosensor for label-free fluorescence detection of proteases. *Small*, 8:3769–3773, 2012. DOI: 10.1002/smll.201201983. 93

[92] Wu, Z., Wang, M., Yang, J., Zheng, X., Cai, W., Meng, G., Qian, H., Wang, H., and Jin, R. Well-defined nanoclusters as fluorescent nanosensors: A case study on $Au_{25}(SG)_{18}$. *Small*, 8:2028–2035, 2012. DOI: 10.1002/smll.201102590. 93, 94, 96

[93] Xia, N., Yang, J., and Wu, Z. Fast, high-yield synthesis of amphiphilic Ag nanoclusters and the sensing of Hg^{2+} in environmental samples. *Nanoscale*, 7:10013–10020, 2015. DOI: 10.1039/c5nr00705d. 93, 94

[94] Wang, S., Meng, X., Feng, Y., Sheng, H., and Zhu, M. An anti-galvanic reduction single-molecule fluorescent probe for detection of Cu(II). *RSC Adv.*, 4:9680–9683, 2014. DOI: 10.1039/c3ra46877a. 93, 94

[95] Muhammed, M. A. H., Verma, P. K., Pal, S. K., Kumar, R. C. A., Paul, S., Omkumar, R. V., and Pradeep, T. Bright, NIR-emitting Au_{23} from Au25: Characterization and applications including biolabeling. *Chem. Eur. J.*, 15:10110–10120, 2009. DOI: 10.1002/chem.200901425. 93, 94

[96] Mathew, A., Sajanlal, P. R., and Pradeep, T. Selective visual detection of TNT at the sub-zeptomole level. *Angew. Chem. Int. Ed.*, 51:9596–9600, 2012. DOI: 10.1002/anie.201203810. 93, 94, 96

[97] George, A., Shibu, E. S., Maliyekkal, S. M., Bootharaju, M. S., and Pradeep, T. Luminescent, freestanding composite films of Au_{15} for specific metal ion sensing. *ACS Appl. Mat. Interfaces*, 4:639–644, 2012. DOI: 10.1021/am201292a. 93, 94

[98] Guan, G., Zhang, S.-Y., Cai, Y., Liu, S., Bharathi, M. S., Low, M., Yu, Y., Xie, J., Zheng, Y., Zhang, Y.-W., and Han, M.-Y. Convenient purification of gold clusters by co-precipitation for improved sensing of hydrogen peroxide, mercury ions and pesticides. *Chem. Commun.*, 50:5703–5705, 2014. DOI: 10.1039/c4cc02008a. 93, 94

[99] Wang, M., Wu, Z., Yang, J., Wang, G., Wang, H., and Cai, W. $Au_{25}(SG)_{18}$ as a fluorescent iodide sensor. *Nanoscale*, 4:4087–4090, 2012. DOI: 10.1039/c2nr30169e. 93, 94, 96

[100] Khatun, E., Ghosh, A., Chakraborty, P., Singh, P., Bodiuzzaman, M., Ganesan, P., Nataranjan, G., Ghosh, J., Pal, S. K., and Pradeep, T. A thirty-fold photoluminescence enhancement induced by secondary ligands in monolayer protected silver clusters. *Nanoscale*, 10:20033–20042, 2018. DOI: 10.1039/c8nr05989f. 93, 94, 97

[101] Yoshimoto, J., Sangsuwan, A., Osaka, I., Yamashita, K., Iwasaki, Y., Inada, M., Arakawa, R., and Kawasaki, H. Optical properties of 2-methacryloyloxyethyl phosphorylcholine-protected Au_4 nanoclusters and their fluorescence sensing of C-reactive protein. *J. Phys. Chem. C*, 119:14319–14325, 2015. DOI: 10.1021/acs.jpcc.5b03934. 93, 95

[102] Gao, X., Lu, Y., Liu, M., He, S., and Chen, W. Sub-nanometer sized $Cu_6(GSH)_3$ clusters: One-step synthesis and electrochemical detection of glucose. *J. Mater. Chem. C*, 3:4050–4056, 2015. DOI: 10.1039/c5tc00246j. 93, 95

[103] Lu, F., Yang, H., Yuan, Z., Nakanishi, T., Lu, C., and He, Y. Highly fluorescent polyethyleneimine protected Au_8 nanoclusters: One-pot synthesis and application in hemoglobin detection. *Sens. Actuators B Chem.*, 291:170–176, 2019. DOI: 10.1016/j.snb.2019.04.067. 93, 95, 97

[104] Zheng, K., Setyawati, M. I., Leong, D. T., and Xie, J. Antimicrobial silver nanomaterials. *Coord. Chem. Rev.*, 357:1–17, 2018. DOI: 10.1016/j.ccr.2017.11.019. 97, 102

[105] Yougbare, S., Chang, T.-K., Tan, S.-H., Kuo, J.-C., Hsu, P.-H., Su, C.-Y., and Kuo, T.-R. Antimicrobial gold nanoclusters: Recent developments and future perspectives. *Int. J. Mol. Sci.*, 20:2924, 2019. DOI: 10.3390/ijms20122924. 97, 102

[106] Tan, X. and Jin, R. Ultrasmall metal nanoclusters for bio-related applications. *WIREs Nanomed. Nanobiotechnol.*, 5:569–581, 2013. DOI: 10.1002/wnan.1237. 98

[107] Makarava, N., Parfenov, A., and Baskakov, I. V. Water-soluble hybrid nanoclusters with extra bright and photostable emissions: A new tool for biological imaging. *Biophys. J.*, 89:572–580, 2005. DOI: 10.1529/biophysj.104.049627. 98

[108] Yu, J., Patel, S. A., and Dickson, R. M. *In vitro* and intracellular production of peptide-encapsulated fluorescent silver nanoclusters. *Angew. Chem. Int. Ed.*, 46:2028–2030, 2007. DOI: 10.1002/anie.200604253. 98

[109] Lin, C.-A. J., Yang, T.-Y., Lee, C.-H., Huang, S. H., Sperling, R. A., Zanella, M., Li, J. K., Shen, J.-L., Wang, H.-H., Yeh, H.-I., Parak, W. J., and Chang, W. H. Synthesis, characterization, and bioconjugation of fluorescent gold nanoclusters toward biological labeling applications. *ACS Nano*, 3:395–401, 2009. DOI: 10.1021/nn800632j. 98, 99

[110] Marjomäki, V., Lahtinen, T., Martikainen, M., Koivisto, J., Malola, S., Salorinne, K., Pettersson, M., and Häkkinen, H. Site-specific targeting of enterovirus capsid by functionalized monodisperse gold nanoclusters. *Proc. Natl. Acad. Sci.*, 111:1277–1281, 2014. DOI: 10.1073/pnas.1310973111. 98

[111] Liu, J., Yu, M., Zhou, C., Yang, S., Ning, X., and Zheng, J. Passive tumor targeting of renal-clearable luminescent gold nanoparticles: Long tumor retention and fast normal tissue clearance. *J. Am. Chem. Soc.*, 135:4978–4981, 2013. DOI: 10.1021/ja401612x. 98, 100

[112] Yu, M., Zhou, J., Du, B., Ning, X., Authement, C., Gandee, L., Kapur, P., Hsieh, J.-T., and Zheng, J. Noninvasive staging of kidney dysfunction enabled by renal-clearable luminescent gold nanoparticles. *Angew. Chem. Int. Ed.*, 55:2787–2791, 2016. DOI: 10.1002/anie.201511148. 98

[113] Yang, J., Xia, N., Wang, X., Liu, X., Xu, A., Wu, Z., and Luo, Z. One-pot one-cluster synthesis of fluorescent and bio-compatible Ag_{14} nanoclusters for cancer cell imaging. *Nanoscale*, 7:18464–18470, 2015. DOI: 10.1039/c5nr06421j. 98, 100

[114] Gan, Z., Lin, Y., Luo, L., Han, G., Liu, W., Liu, Z., Yao, C., Weng, L., Liao, L., Chen, J., Liu, X., Luo, Y., Wang, C., Wei, S., and Wu, Z. Fluorescent gold nanoclusters with interlocked staples and a fully thiolate-bound kernel. *Angew. Chem. Int. Ed.*, 55:11567–11571, 2016. DOI: 10.1002/anie.201606661. 98

[115] Wang, S., Meng, X., Das, A., Li, T., Song, Y., Cao, T., Zhu, X., Zhu, M., and Jin, R. A 200—fold quantum yield boost in the photoluminescence of silver—doped $Ag_x Au_{25-x}$ nanoclusters: The 13th silver atom matters. *Angew. Chem. Int. Ed.*, 53:2376–2380, 2014. DOI: 10.1002/anie.201307480. 99

[116] Hu, D., Sheng, Z., Fang, S., Wang, Y., Gao, D., Zhang, P., Gong, P., Ma, Y., and Cai, L. Folate receptor-targeting gold nanoclusters as fluorescence enzyme mimetic nanoprobes for tumor molecular colocalization diagnosis. *Theranostics*, 4:142–153, 2014. DOI: 10.7150/thno.7266. 99, 101

[117] Kurdekar, A. D., Avinash Chunduri, L. A., Manohar, C. S., Haleyurgirisetty, M. K., Hewlett, I. K., and Venkataramaniah, K. Streptavidin-conjugated gold nanoclusters as ultrasensitive fluorescent sensors for early diagnosis of HIV infection. *Sci. Adv.*, 4:eaar6280, 2018. DOI: 10.1126/sciadv.aar6280. 101

[118] Loynachan, C. N., Soleimany, A. P., Dudani, J. S., Lin, Y., Najer, A., Bekdemir, A., Chen, Q., Bhatia, S. N., and Stevens, M. M. Renal clearable catalytic gold nanoclusters for in vivo disease monitoring. *Nat. Nanotechnol.*, 14:883–890, 2019. DOI: 10.1038/s41565-019-0527-6. 101

[119] Wang, X., Gao, W., Xu, W., and Xu, S. Fluorescent Ag nanoclusters templated by carboxymethyl-β-cyclodextrin (CM-β-CD) and their *in vitro* antimicrobial activity. *Mater. Sci. Eng. C*, 33:656–662, 2013. DOI: 10.1016/j.msec.2012.10.012. 102, 103

[120] Yuan, X., Setyawati, M. I., Tan, A. S., Ong, C. N., Leong, D. T., and Xie, J. Highly luminescent silver nanoclusters with tunable emissions: Cyclic reduction-decomposition synthesis and antimicrobial properties. *NPG Asia Mater.*, 5:e39-e39, 2013. DOI: 10.1038/am.2013.3. 102, 103

[121] Zheng, K., Setyawati, M. I., Leong, D. T., and Xie, J. Surface ligand chemistry of gold nanoclusters determines their antimicrobial ability. *Chem. Mater.*, 30:2800–2808, 2018. DOI: 10.1021/acs.chemmater.8b00667. 102

[122] Zheng, K., Li, K., Chang, T.-H., Xie, J., and Chen, P.-Y. Synergistic antimicrobial capability of magnetically oriented graphene oxide conjugated with gold nanoclusters. *Adv. Funct. Mater.*, 29:1904603, 2019. DOI: 10.1002/adfm.201904603. 102

[123] Qing, Z., Mingying, Y., Ye, Z., and Chuanbin, M. Metallic nanoclusters for cancer imaging and therapy. *Curr. Med. Chem.*, 25:1379–1396, 2018. DOI: 10.2174/0929867324666170331122757. 102

[124] Du, B., Jiang, X., Das, A., Zhou, Q., Yu, M., Jin, R., and Zheng, J. Glomerular barrier behaves as an atomically precise bandpass filter in a sub-nanometre regime. *Nat. Nanotechnol.*, 12:1096–1102, 2017. DOI: 10.1038/nnano.2017.170. 102

[125] Jiang, X., Du, B., and Zheng, J. Glutathione-mediated biotransformation in the liver modulates nanoparticle transport. *Nat. Nanotechnol.*, 14:874–882, 2019. DOI: 10.1038/s41565-019-0499-6. 104

[126] Zhang, X.-D., Luo, Z., Chen, J., Shen, X., Song, S., Sun, Y., Fan, S., Fan, F., Leong, D. T., and Xie, J. Ultrasmall $Au_{10-12}(SG)_{10-12}$ nanomolecules for high tumor specificity and cancer radiotherapy. *Adv. Mater.*, 26:4565–4568, 2014. DOI: 10.1002/adma.201400866. 104

[127] Zhang, X.-D., Luo, Z., Chen, J., Song, S., Yuan, X., Shen, X., Wang, H., Sun, Y., Gao, K., Zhang, L., Fan, S., Leong, D. T., Guo, M., and Xie, J. Ultrasmall glutathione-protected gold nanoclusters as next generation radiotherapy sensitizers with high tumor uptake and high renal clearance. *Sci. Rep.*, 5:8669, 2015. DOI: 10.1038/srep08669. 104

[128] He, F., Yang, G., Yang, P., Yu, Y., Lv, R., Li, C., Dai, Y., Gai, S., and Lin, J. A new single 808 nm NIR light-induced imaging-guided multifunctional cancer therapy platform. *Adv. Funct. Mater.*, 25:3966–3976, 2015. DOI: 10.1002/adfm.201500464. 104, 105

[129] Lei, Y., Tang, L., Xie, Y., Xianyu, Y., Zhang, L., Wang, P., Hamada, Y., Jiang, K., Zheng, W., and Jiang, X. Gold nanoclusters-assisted delivery of NGF siRNA for effective treatment of pancreatic cancer. *Nat. Commun.*, 8:15130, 2017. DOI: 10.1038/ncomms15130. 105

[130] Yang, D., Gulzar, A., Yang, G., Gai, S., He, F., Dai, Y., Zhong, C., and Yang, P. Au nanoclusters sensitized black TiO_{2-x} nanotubes for enhanced photodynamic therapy driven by near-infrared light. *Small*, 13:1703007, 2017. DOI: 10.1002/smll.201703007. 106

[131] Katla, S. K., Zhang, J., Castro, E., Bernal, R. A., and Li, X. Atomically precise $Au_{25}(SG)_{18}$ nanoclusters: Rapid single-step synthesis and application in photothermal therapy. *ACS Appl. Mat. Interfaces*, 10:75–82, 2018. DOI: 10.1021/acsami.7b12614. 106

[132] Greeley, J., Jaramillo, T. F., Bonde, J., Chorkendorff, I., and Nørskov, J. K. Computational high-throughput screening of electrocatalytic materials for hydrogen evolution. *Nat. Mater.*, 5:909–913, 2006. DOI: 10.1038/nmat1752. 106

[133] Zeng, K. and Zhang, D. Recent progress in alkaline water electrolysis for hydrogen production and applications. *Prog. Energy Combust. Sci.*, 36:307–326, 2010. DOI: 10.1016/j.pecs.2009.11.002. 106

[134] Zhao, S., Jin, R., Song, Y., Zhang, H., House, S. D., Yang, J. C., and Jin, R. Atomically precise gold nanoclusters accelerate hydrogen evolution over MoS_2 nanosheets: The dual interfacial effect. *Small*, 13:1701519, 2017. DOI: 10.1002/smll.201701519. 106, 107

[135] Zhao, S., Jin, R., Abroshan, H., Zeng, C., Zhang, H., House, S. D., Gottlieb, E., Kim, H. J., Yang, J. C., and Jin, R. Gold nanoclusters promote electrocatalytic water oxidation at the nanocluster/$CoSe_2$ interface. *J. Am. Chem. Soc.*, 139:1077–1080, 2017. DOI: 10.1021/jacs.6b12529. 106, 108

[136] Sakai, N. and Tatsuma, T. Photovoltaic properties of glutathione-protected gold clusters adsorbed on TiO_2 electrodes. *Adv. Mater.*, 22:3185–3188, 2010. DOI: 10.1002/adma.200904317. 106

[137] Zhang, D., Choy, W. C. H., Wang, C. C. D., Li, X., Fan, L., Wang, K., and Zhu, H. Polymer solar cells with gold nanoclusters decorated multi-layer graphene as transparent electrode. *Appl. Phys. Lett.*, 99:223302, 2011. DOI: 10.1063/1.3664120. 106

[138] Chen, Y.-S., Choi, H., and Kamat, P. V. Metal-cluster-sensitized solar cells. A new class of thiolated gold sensitizers delivering efficiency greater than 2%. *J. Am. Chem. Soc.*, 135:8822–8825, 2013. DOI: 10.1021/ja403807f. 106, 109

[139] Choi, H., Chen, Y.-S., Stamplecoskie, K. G., and Kamat, P. V. Boosting the photovoltage of dye-sensitized solar cells with thiolated gold nanoclusters. *J. Phys. Chem. Lett.*, 6:217–223, 2015. DOI: 10.1021/jz502485w. 107, 110

[140] Abbas, M. A., Kim, T.-Y., Lee, S. U., Kang, Y. S., and Bang, J. H. Exploring interfacial events in gold-nanocluster-sensitized solar cells: Insights into the effects of the cluster size and electrolyte on solar cell performance. *J. Am. Chem. Soc.*, 138:390–401, 2016. DOI: 10.1021/jacs.5b11174. 107, 110

[141] Yuan, P., Chen, R., Zhang, X., Chen, F., Yan, J., Sun, C., Ou, D., Peng, J., Lin, S., Tang, Z., Teo, B. K., Zheng, L.-S., and Zheng, N. Ether-soluble Cu53 nanoclusters as an effective precursor of high-quality CuI films for optoelectronic applications. *Angew. Chem. Int. Ed.*, 58:835–839, 2019. DOI: 10.1002/anie.201812236. 108

[142] Niesen, B. and Rand, B. P. Thin film metal nanocluster light-emitting devices. *Adv. Mater.*, 26:1446–1449, 2014. DOI: 10.1002/adma.201304725. 109

[143] Koh, T. W., Hiszpanski, A. M., Sezen, M., Naim, A., Galfsky, T., Trivedi, A., Loo, Y. L., Menon, V., and Rand, B. P. Metal nanocluster light-emitting devices with suppressed parasitic emission and improved efficiency: Exploring the impact of photophysical properties. *Nanoscale*, 7:9140–9146, 2015. DOI: 10.1039/c5nr01332a. 109

[144] Nishikitani, Y., Takizawa, D., Uchida, S., Lu, Y., Nishimura, S., Oyaizu, K., and Nishide, H. Ag nanocluster-based color converters for white organic light-emitting devices. *J. Appl. Phys.*, 122:184302, 2017. DOI: 10.1063/1.4995671. 109, 111

[145] Wang, Z., Chen, B., Susha, A. S., Wang, W., Reckmeier, C. J., Chen, R., Zhong, H., and Rogach, A. L. All-copper nanocluster based down-conversion white light-emitting devices. *Adv. Sci.*, 3:1600182, 2016. DOI: 10.1002/advs.201600182. 110

[146] Zhang, H., Liu, H., Tian, Z., Lu, D., Yu, Y., Cestellos-Blanco, S., Sakimoto, K. K., and Yang, P. Bacteria photosensitized by intracellular gold nanoclusters for solar fuel production. *Nat. Nanotechnol.*, 13:900–905, 2018. DOI: 10.1038/s41565-018-0267-z. 111

CHAPTER 5

Summary and Outlook

The previous chapters introduced the emerging field of atomically precise metal nanocluster research. In retrospect, the research on metal nanoparticles has experienced a long track, from the early stage of polydisperse NPs in the 19th century, to the achievement of monodisperse ones by the early 21st century, and now finally to the atomically precise nanoclusters. Currently, atomic precision has been achieved in the 1–3 nm size regime (tens to hundreds of atoms). The establishment of the "size-focusing" synthetic methodology has played a major role, and mechanistic understanding (i.e., the "kinetic control and thermodynamic selection" principle) led to the generalization of the methodology and attainment of a series of atomically precise gold nanoclusters, including some other metals such as silver and alloy nanoclusters. Together with the post-synthesis conversion strategies, all these have greatly contributed to the establishment and flourish of this field. Further understanding of the synthesis principles and the development of novel synthesis methods are still needed, especially new strategies for site-specific tailoring of nanoclusters, which will certainly provide new opportunities in future research on metal nanoclusters.

The characterizations of atomically precise metal nanoclusters, including the size and structure determination, composition analysis, and studies on the optical, electrochemical, and magnetic properties, are crucial not only for understanding the new properties of nanocluster materials and obtaining the structure-property relationships, but also for the development of applications of nanoclusters in biomedical, environmental, energy, and catalysis fields. In the future, new characterization methods are still needed, in particular how to monitor the growth mechanism in situ and in real time, as well as the catalytic process. In the latter, rich dynamics would be occurring at the catalytic active-sites, but new techniques with atomic spatial resolution and millisecond to microsecond temporal resolution are critically needed. Such new techniques will certainly bring the nanocluster research to another exciting level.

To tackle larger NCs with atomic precision (e.g., toward a thousand and even many thousands of atoms in the metal core), new breakthroughs are still needed. The precise atomic composition determination by mass spectrometry is currently still difficult for more than 500 metal atoms, and the total structure determination of giant NCs still have major issues of insufficient crystal quality, weak diffraction by surface ligands, and some other challenges. But we believe future research will make more exciting progress toward atomically precise nanoclusters.

As a new class of nanomaterials, atomically precise metal nanoclusters are expected to have a great impact on the fundamental understanding of the structure-property relationships at the ultimate atomic level. Based upon such new knowledge, atomically precise nanoclusters will find new opportunities in both fundamental research and practical applications.

Authors' Biographies

ZHIKUN WU

Zhikun Wu is a Professor of Chemistry at the Institute of Solid State Physics, Chinese Academy of Sciences. He received his Ph.D. in Chemistry from the Institute of Chemistry, Chinese Academy of Sciences in 2004. His current research interests are the syntheses, structures, properties and applications of atomically precise metal nanoclusters.

RONGCHAO JIN

Rongchao Jin is a Professor of Chemistry at Carnegie Mellon University. He received his B.S. in Chemical Physics from the University of Science and Technology of China in 1995, his M.S. in Catalysis from Dalian Institute of Chemical Physics in 1998, and his Ph.D. in Chemistry from Northwestern University in 2003. After three years of postdoctoral research at the University of Chicago, he joined the chemistry faculty of CMU in 2006 and was promoted to Associate Professor in 2012 and Full Professor in 2015. His current research interests are atomically precise metal nanoparticles, nano-optics, and catalysis.

Printed in the United States
by Baker & Taylor Publisher Services